SPRINKLER IRRIGATION

Dr. S.N. Yadav, presently Associate Professor at College of Agricultural Engineering and Post Harvest Technology (Central Agricultural University) Sikkim. He has 26 years of vast experience in the field of Agricultural Engineering while working in Indian Council of Agricultural Research (ICAR). Out of 26 years, he has served for 16 years at Central Institute of Agricultural Engineering, Bhopal.

He has more than 10 research papers, popular articles, 12 manuals, 2 extension bulletins in his credit. Besides various technical reports, he has edited 2 books related to agricultural engineering.

SPRINKLER IRRIGATION

— *Editor* —

Dr. S.N. Yadav

Associate Professor

College of Agricultural Engineering and Post Harvest Technology

Central Agricultural University

Sikkim

BIOTECH

2020

BIOTECH BOOKS®

ISBN 81-7622-223-2
ISBN 978-81-7622-223-5

Published by : **BIOTECH BOOKS**
4762-63/23, Ansari Road, Darya Ganj,
New Delhi - 110 002
Ph.: 011 43003222
e-mail: biotechbooks@yahoo.co.in

Laser Typesetting : **Classic Computer Services**
Delhi - 110 035

Printed at : **Chawla Offset Printers**
Delhi - 110 052

PRINTED IN INDIA

Preface

Sprinkler irrigation is the method of applying irrigation water which is similar to rainfall. Water is distributed through a system of pipes usually by pumping. It is then sprayed into the air and irrigates the entire soil surface through spray heads so that it breaks up into small water drops which fall to the ground. This methodology of irrigation, meant to reduce irrigation time is an ideal means of providing efficient coverage for any field size, apart from being adaptable to nearly all irrigable soils, since sprinklers are available in a wide range of discharge capacity.

The present text aims to comprehend the techniques and mechanisms which entail sprinkler irrigation, making sure to acquaint readers with the relevance of this irrigation technology. Delineating the various methods through which sprinkler irrigation is executed, it also undertakes a comprehensive look at the numerous kinds of products that are available in the market to achieve this, guiding readers on the best possible machinery. In addition, relevant information on current trends and developments in the field have been added to the book's content, with insight on the changing methods associated with sprinkler irrigations.

S.N. Yadav

Preface

Contents

1

Introduction

Sprinkler irrigation is a method of applying irrigation water which is similar to rainfall. Water is distributed through a system of pipes usually by pumping. It is then sprayed into the air and irrigated entire soil surface through spray heads so that it breaks up into small water drops which fall to the ground.

Sprinklers provide efficient coverage for small to large areas and are suitable for use on all types of soil terrain and conditions. It is adaptable to nearly all soils since sprinklers are available in a wide range of discharge capacity. Irrigation sprinklers are used on farms, golf courses, residential yards, and other places to water crops, lawns, gardens or other plants in the event of drought. They are also used for recreation or as a cooling system.

Sprinkler Types

Industrial

Sprinklers that spray in a fixed pattern are generally called sprays or spray heads. Sprays are not usually designed to operate at pressures above 200 kPa, due to misting problems that may develop.

Higher pressure sprinklers that rotate around themselves are driven by a ball drive, gear drive, or impact mechanism (impact sprinklers). These can be designed to rotate in a full or partial circle.

Some sprinklers are also known as Floppy Sprinkler, Spray Pop-ups, pulsating sprinklers due to their water stream and revolutionary new concept having no rotating or moving parts.

Raingun

Rainguns are similar to impact sprinkler, except that they generally operate at very high pressures of 275 to 900 kPa and flows of 3 to 76 L/s, usually with nozzle diameters in the range of 10 to 50 mm. In addition to irrigation, guns are used for industrial applications such as dust suppression and logging.

Many irrigation sprinklers are buried in the ground along with their supporting plumbing, but above ground and moving sprinklers are also common. Most irrigation sprinklers are functioned through electric and hydraulic technology and are grouped together in zones that can be collectively turned on and off by actuating a solenoid-controlled valve.

Underground Sprinklers

Underground sprinklers function through means of basic electronic and hydraulic technology. A 110 volt supplied controller, or sprinkler box sends a pre-determind 24 volt signal to an awaiting solenoid valve buried in the yard. This valve and all of the sprinklers that will be activated by this valve are known as a zone. Upon activation, the solenoid, which sits on top of the valve is magnetised lifting a small stainless steel plunger in its center. By doing this, the activated (or raised) plunger allows air to escape from the top of a rubber diaphragm located in the center of the valve. Water that has been charged and waiting on the bottom of this same diaphragm now has the higher pressure and lifts the diaphragm. This pressurised water is then allowed to escape down stream of the valve through a series of PVC pipes. Along these pipes and flush to grade level are pre measured and spaced out sprinklers. These sprinklers can be fixed spray heads that have a set pattern and generally spray between 2 and 5 m, full rotating sprinklers that can spray a broken stream of water from 6 to 12 feet, or small drip emitters that release a slow, steady drip of water on more delicate plants such as flowers and shrubs.

Sprinkler Use

Most irrigation sprinklers are used as part of a sprinkler system, consisting of various plumbing parts, piping and control equipment. Piping is connected to the water source via plumbing fittings and the control system opens and closes valves to provide water on a schedule. The control provided varies depending on the equipment used; some systems are fully automated and even compensate for rain, runoff and evaporation, while others require much more user attention for the same effectiveness. The pump supply system, sprinklers and operating conditions must be designed to enable a uniform application of water.

Suitable Crops

Sprinkler irrigation is suited for most row, field and tree crops and water can be sprayed over or under the crop canopy. However, large sprinklers are not recommended for irrigation of delicate crops because the large water drops produced by the sprinklers may damage the crop.

Suitable Slopes

Sprinkler irrigation is adaptable to any farmable slope, whether uniform or undulating. The lateral pipes supplying water to the sprinklers should always be laid out along the land contour whenever possible. This will minimise the pressure changes at the sprinklers and provide a uniform irrigation.

Suitable Soils

Sprinklers are best suited to sandy soils with high infiltration rates although they are adaptable to most soils. The average application rate from the sprinklers (in mm/hour) is always chosen to be less than the basic infiltration rate of the soil so that surface ponding and runoff can be avoided.

Sprinklers are not suitable for soils which easily form a crust. If sprinkler irrigation is the only method available, then light fine sprays should be used. The larger sprinklers producing larger water droplets are to be avoided.

Suitable Irrigation Water

A good clean supply of water, free of suspended sediments, is required to avoid problems of sprinkler nozzle blockage and spoiling the crop by coating it with sediment.

Components of Sprinkler System

A typical sprinkler irrigation system consists of the following components:

— Pump unit

— Mainline and sometimes submainlines

— Laterals

— Sprinklers

The pump unit is usually a centrifugal pump which takes water from the source and provides adequate pressure for delivery into the pipe system.

The mainline - and submainlines - are pipes which deliver water from the pump to the laterals. In some cases these pipelines are permanent and are laid on the soil surface or buried below ground. In other cases they are temporary, and can be moved from field to field. The main pipe materials used include asbestos cement, plastic or aluminium alloy.

The laterals deliver water from the mainlines or submainlines to the sprinklers. They can be permanent but more often they are portable and made of aluminium alloy or plastic so that they can be moved easily.

It consists of a system of lightweight aluminium or plastic pipes which are moved by hand. The rotary sprinklers are usually spaced 9-24 m apart along the lateral which is normally 5-12.5 cm in diameter. This is so it can be carried easily. The lateral pipe is located in the field until the irrigation is complete. The pump is then switched off and the lateral is disconnected from the mainline and moved to the next location. It is re-assembled and connected to the mainline and the irrigation begins again. The lateral can be moved one to four times a day. It is gradually moved around the field until the whole field is irrigated. This is the simplest of all systems. Some use more than one lateral to irrigate larger areas.

A common problem with sprinkler irrigation is the large labour force needed to move the pipes and sprinklers around the field. In some places such labour may not be available and may also be costly. To overcome this problem many mobile systems have been developed such as the hose reel raingun and the centre pivot. However, these systems go beyond the scope of this introduction to sprinkler systems. Another system which does not need a large labour force is the drag-hose sprinkler system. Main and laterals are buried PVC pipes: one lateral covers three positions.

Operating Sprinkler Systems

The main objective of a sprinkler system is to apply water as uniformly as possible to fill the root zone of the crop with water.

Wetting Patterns

The wetting pattern from a single rotary sprinkler is not very uniform (Figure 1). Normally the area wetted is circular (topview). The heaviest wetting is close to the sprinkler. For good uniformity several sprinklers must be operated close together so that their patterns overlap. For good uniformity the overlap should be at least 65% of the wetted diameter. This determines the maximum spacing between sprinklers.

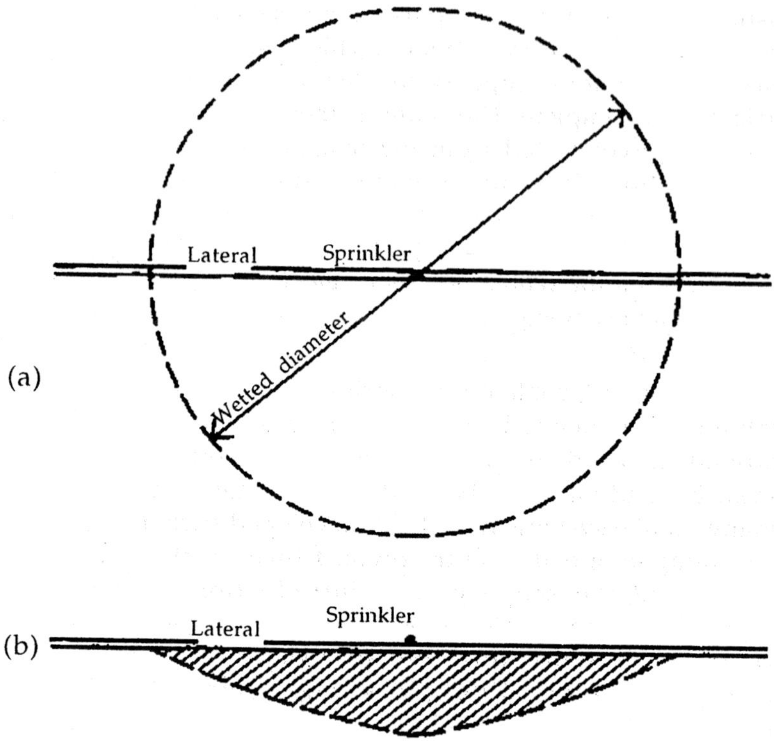

Figure 1: (a) Wetting pattern for a single sprinkler (Top view) b) (Side view)

The uniformity of sprinkler applications can be affected by wind and water pressure.

Spray from sprinklers is easily blown about by even a gentle breeze and this can seriously reduce uniformity. To reduce the effects of wind the sprinklers can be positioned more closely together.

Sprinklers will only work well at the right operating pressure recommended by the manufacturer. If the pressure is above or below this then the distribution will be affected. The most common problem is when the pressure is too low. This happens when pumps and pipes wear. Friction increases and so pressure at the sprinkler reduces. The result is that the water jet does not break up and all the water tends to fall in one area towards the outside of the wetted circle. If the pressure is too high then the distribution will also be poor. A fine spray develops which falls close to the sprinkler.

Application Rate

This is the average rate at which water is sprayed onto the crops and is measured in mm/hour. The application rate depends on the size of sprinkler nozzles, the operating pressure and the distance between sprinklers. When selecting a sprinkler system it is important to make sure that the average application rate is less than the basic infiltration rate of the soil. In this way all the water applied will be readily absorbed by the soil and there should be no runoff.

Sprinkler Drop Sizes

As water sprays from a sprinkler it breaks up into small drops between 0.5 and 4.0 mm in size. The small drops fall close to the sprinkler whereas the larger ones fall close to the edge of the wetted circle. Large drops can damage delicate crops and soils and so in such conditions it is best to use the smaller sprinklers.

Drop size is also controlled by pressure and nozzle size. When the pressure is low, drops tend to be much larger as the water jet does not break up easily. So to avoid crop and soil damage use small diameter nozzles operating at or above the normal recommended operating pressure.

Selecting a Sprinkler Irrigation System

If the sprinkler system is for a new installation, you have two important tasks you must perform before purchasing the system. First, you must check the county soil survey maps to make sure the soils in the field are irrigable. Second, you must have a readily available source of water near the field and have a State Water Commission-issued water permit for that water. The water source must be of sufficient quantity and quality for successful irrigation.

A sprinkler "throws" water through the air in an effort to simulate rainfall whereas the other three irrigation methods apply water directly to the soil, either on or below the surface. A sprinkler system can be composed of one or many sprinklers. In systems that use many sprinklers, the sprinklers are attached to a pipeline at a predetermined spacing in order to achieve a uniform application amount. When selecting a sprinkler system, the most important *physical* parameters to consider are:

1. The shape and size (acres) of the field.
2. The topography of the field. Are there many hills with steep slopes?
3. The amount of time and labor required to operate the system. How much time and labor do you have available?

The center pivot system is very adaptable but doesn't work very well on irregularly shaped fields, long nar-row fields and fields that contain some type of obstruction (trees, farmsteads, etc.). In these situations, other sprinkler systems may be more effective.

Sprinkler System Capacity

The sprinkler system capacity is the flow rate the system needs to irrigate an area adequately. It is expressed in gallons per minute per acre (gpm/acre). The system capacity is dependent on the:

1. Peak crop water requirements during the growing season.
2. Effective crop rooting depth.
3. Texture and infiltration rate of the soil.
4. The available water holding capacity of the soil.
5. Pumping capacity of the well or wells (if wells are the water source).
6. The State Water Commission permitted pumping rate.

Table 1 shows the system capacity you need for the most common irrigated crops in North Dakota and various soil textures. To use this table, you must determine the dominant soil texture in the field and what type of crops you will grow (the crop rotation), then determine the appropriate system capacity.

Table 1. System Capacity in gallons per minute per acre (gpm/acre) for different soil textures needed to supply sufficient water for each crop in 9 out of 10 years. An application efficiency of 80% and a 50% depletion of available soil water were used for the calculations.

Crop	*Root Zone Depth (ft)* *Coarse Sand and Gravel* *Sand* *Loamy Sand* *Sandy Loam Fine Sandy Loam* *Loam and Silt Loam*
Potatoes	2.0
	8.2
	7.5
	7.0
	6.4
	6.1
	5.7
Dry Beans	2.0
	7.9
	7.1
	6.4
	6.1
	5.7
	5.4
Soybeans	2.0
	7.9
	7.1
	6.4
	6.1
	5.7
	5.4
Corn	3.0

	7.3
	6.6
	5.9
	5.5
	5.3
	4.9
Sugarbeets	3.0
	7.3
	6.6
	5.9
	5.5
	5.3
	4.9
Small Grains	3.0
	7.3
	6.6
	5.9
	5.5
	5.3
	4.9
Alfalfa	4.0
	6.8
	5.9
	5.6
	5.1
	5.0
	4.5

For example, if you plan a rotation of potatoes, corn and alfalfa on a loamy sand, you can determine from Table 1 that potatoes require 7.0 gpm/acre, corn 5.9 gpm/acre and alfalfa 5.6 gpm/acre. You would select a design system capacity for the crop requiring the largest amount, in this case the potatoes at 7.0 gpm/acre. If you install a center pivot system covering 130 acres, you would need about 910 gpm for proper design. However, what you need for proper design and what a well will produce is frequently different. As a general rule, under full season irrigation, you need a minimum of 6 gpm/acre is required because most of the soils irrigated in North Dakota are loamy sands or sandy loams. You can use a lesser flow rate, but you will need to do more intensive water management.

A sprinkler system must be designed to apply water uniformly without runoff or erosion. The application rate of the sprinkler system must match to the intake rate of the most restrictive soil in the field. If the application rate exceeds the soil intake rate, the water will run off the field or relocate within the field resulting in over- and underwatered areas. Using tillage that improves surface storage, such as deep cultivation or making basins, will help control runoff.

Selecting the Appropriate Sprinkler Systems

There are five most common sprinkler systems:

1. A square 160 acre field.
2. A 100 feet deep well near the center of the field.
3. The water supply is adequate for any sprinkler system.
4. The soils are suitable for the system application rate.

Center Pivot

This self-propelled sprinkler system rotates around the pivot point and has the lowest labor requirements of the systems considered. It is constructed using a span of pipe connected to moveable towers. It will irrigate approximately 130 acres out of a square quarter section.

Center pivot systems are either electric, water, or oil-drive and can handle slopes up to 15 percent. Sprinkler packages are available for low to high operating pressures (25 to 80 psi at the pivot point). Sprinklers can be mounted on top of the spans or on drop-tubes which put them closer to the crop. The speed of the rotation controls the water application amount. Center pivots are adaptable for any height crop and are particularly suited to lighter soils. They are generally not recommended for heavy soils with low infiltration rates.

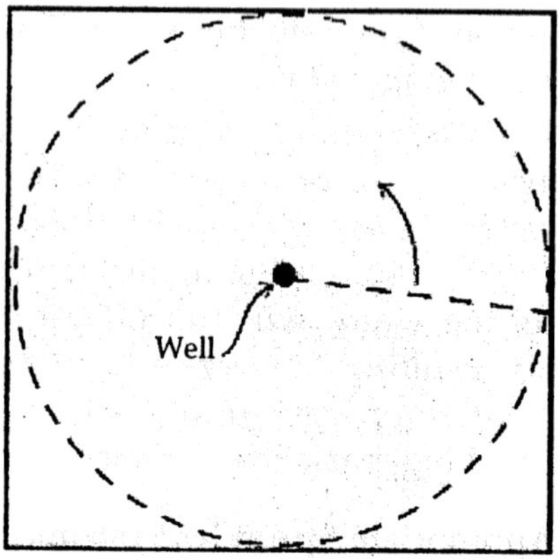

Deep wheel tracks can be a problem on some soils, but a number of management methods are available to control this problem. Electric-drive pivots are the most popular due to flexibility of operation. Computerised control panels allow the operator to specify speed changes at any place in the field, reverse the pivot, turn on auxiliary pumps at a specified time and many other features.

Center Pivot With Corner Attachment

Corner attachment systems are available which allow irrigation of most of the corner areas that a conventional center pivot system misses. Depending on the method of corner irrigation, pivot systems with corner attachments will irrigate 145 to 152 acres out of a 160 acre quarter section. The most common method of corner irrigation has an additional span, complete with tower, attached to the end of the center pivot system mainline which swings out in the corners. As it swings out, sprinklers turn on to irrigate the corners Either a buried wire or mechanical switch controls the movement of the moving span.

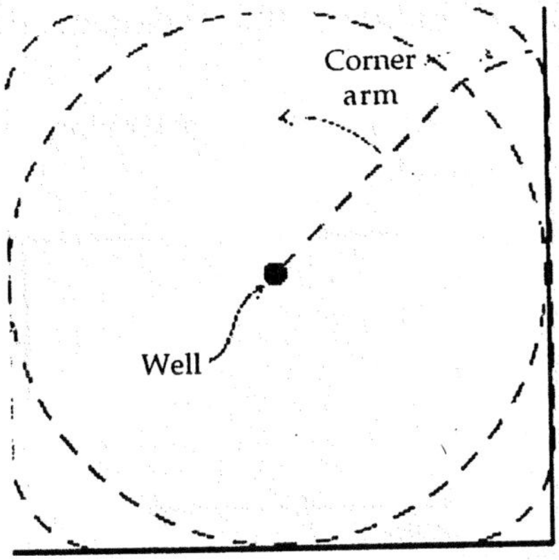

Another type of corner system uses several guns mounted on the end of the center pivot system mainline. The system activates the guns in sequence from smallest to largest and back again as the machine moves past the corners. A corner span generally costs about half as much as the rest of the pivot, increasing the capital cost per acre on a square 160 acres. However, if the field is rectangular, the corner span can be extended on one or both ends, increasing the amount of irrigated acreage from 170 to 185 acres. High value crops and/or high land value as well as scarcity of irrigable land is necessary to justify additional costs over a "plain" center pivot.

Linear Move

The linear move (sometimes called a lateral move) irrigation system is built the same way as a center pivot, with moving towers and spans of pipe connecting the towers. The main difference is that all the towers move at the same speed and in the same direction. Water is pumped into either one of the ends or into the center. Water can be supplied to the linear system either through a canal, by dragging a supply hose which is connected to a mainline, or by connecting and disconnecting from hydrants as the system moves down the field. To gain acreage and make the transition from one side of the field to the other, some linear systems pivot at the end of the field.

The lateral movement makes powering a linear system with electricity difficult.

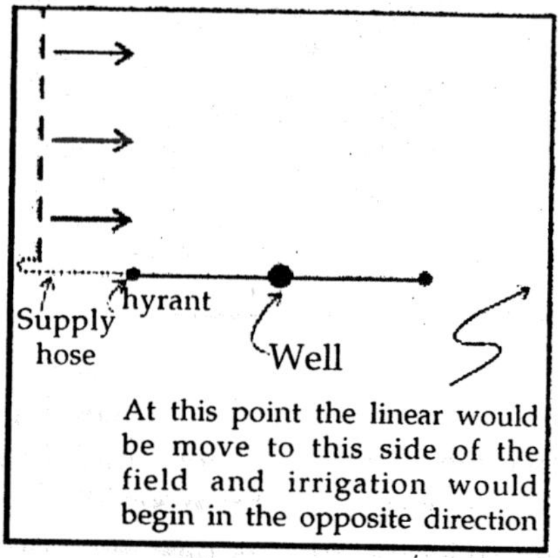

The primary advantage of the linear is that it can irrigate rectangular fields up to a mile long and a half mile wide. Due to the high capital investment cost, linear systems are used on high value crops such as potatoes, vegetables and turf.

Traveling Big Gun

The traveling big gun system uses a large capacity nozzle (3/4 to 1 1/2 inches in diameter) and high pressure (90 to 125 psi) to throw water out over the crop (175 to 350 foot radius) as it is pulled through an alley in the field. Traveling big guns come in two main configurations: hard hose or flexible hose feed. With the hard hose system, a hard polyethylene hose is wrapped on a reel mounted on a trailer. The trailer is anchored at the end or center of the field. The gun is connected to the end of the hose and is pulled to the end of the field. The hose wrapping up on the reel pulls the gun across the field.

With the flexible hose system, the gun is mounted on a four-wheel cart. A flexible hose from the mainline supplies water to the gun. A winch cable on the cart pulls the cart through the field. The cable is anchored at the end of the field. Most traveling big gun systems have their own power unit and cable winch mounted directly on the machine. The power unit may be an internal combustion engine or a water drive.

Particularly adaptable to various crop heights, variable travel speeds, odd shaped fields, and rough terrain, the big gun requires a moderate initial investment, more labor and higher operating pressures than center pivots and linear systems. One 1,320-foot (quarter-mile) long set usually covers 8 to 10 acres, but many variations are available using different water quantities and operating pressures. This system sacrifices irrigated cropland because the alley generally is two rows wide. Most big gun systems are used on a maximum of 80 to 100 acres per gun.

Side Roll System

The side roll (sometimes called a wheel roll) system, as shown, consists of a lateral, usually one-quarter mile long, mounted on 4- to 10-foot diameter wheels with the pipe acting as an axle. Common pipe diameters are 4 and 5 inches. The side roll irrigates an area from 60 to 90 feet wide. When the system has applied the desired amount or water to this set area, a gasoline engine located at the center moves the side roll to the next set. The sprinklers generally are mounted on weighted, swivelling connectors so that no matter where the side roll stops, the sprinklers will always be right side up. This type of system is not recommended for slopes greater than 5 percent and should be used mainly on flat ground. When not being used, side rolls are subject to damage from high winds.

Other characteristics include: adapted only to low-growing crops, medium labor requirements, moderate initial investment, medium operating pressure (50 psi at inlet), generally rectangular field requirements, and each lateral capable of irrigating a maximum of 40 acres.

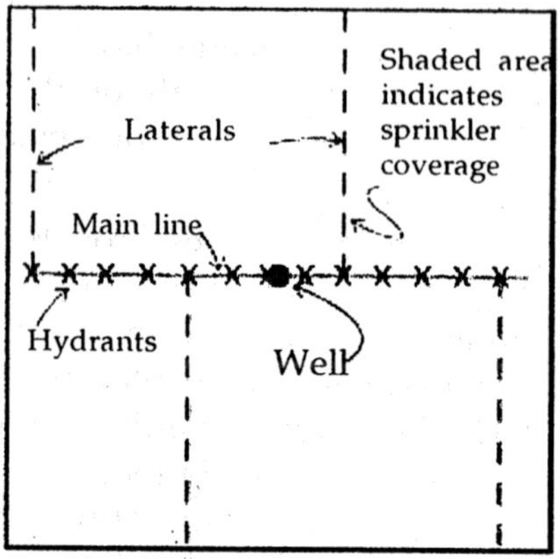

The side roll is better adapted to heavier soils than a continuous moving system. You must purchase special wheels for moving this system from field to field without disassembly. One variation of the side roll system has trail lines with up to three additional sprinklers on a 60-foot spacing. This reduces the number of sets required to irrigate a particular field.

References

Abdi Qassim, *Sprinkler Irrigation, Situation Analyse,* IPTRID, 2003.

Thomas F. Scherer at all, *Sprinkler Irrigation Systems,* MWPS, 1999.

2

Sprinkler Irrigation Planning

A sprinkler irrigation system generally includes sprinklers, laterals, submains, main pipelines, pumping plants and boosters, operational control equipment and other accessories required for efficient water application. In some cases, sprinkler systems may be pressurised by gravity and therefore pumping plants may not be required.

The planning and design of irrigation systems should aim at maximising the returns and minimising both the initial capital outlay and the costs per unit volume of water used, thus contributing both directly and indirectly to the overall reduction of the production costs and the increase of returns. In other words, planning and design is a process of optimising resources. The procedure for designing sprinkler systems can be divided into two phases:

1. Preliminary design steps
2. Adjustment or final design steps

Preliminary design steps comprise the procedure for synthesising farm data in order to determine preliminary design parameters, which will be needed in the final design adjustment process. The final design steps reconcile the preliminary design parameters obtained with the irrigation equipment performance characteristics, as well as human, physical and financial factors. In fact, the final adjustment of the design is the process of selecting the appropriate irrigation system components for the specific circumstances.

They are put in place to protect the fields from flooding through high intensity rainstorms. As not all sprinkler irrigation systems can be covered within the scope of the manual, the following sprinkler irrigation systems will be used as examples to illustrate the design procedure:

1. Semi-portable sprinkler irrigation system for an individual farm
2. Semi-portable sprinkler irrigation system for a smallholder scheme (system for several small plot holders)
3. Drag-hose sprinkler irrigation system for a smallholder scheme (system for several small plot holders)
4. Hose-drag travelling irrigator for individual farm
5. Hose-pull travelling irrigator for individual farm.

The same type of field with the same contour lines will be used for the design of the first three systems and another type of field for the design of the last two continuous-move or travelling sprinkler irrigators. The outputs of the designs are alternative irrigation system options for possible adoption. Once the components of each system are selected, a bill of quantities will be drawn up for each case in order to estimate the cost of the project.

Principles of Preliminary Design

The first step in the preliminary design phase is the collection of basic farm data. The data include:

- a topographic map showing:
  - The proposed irrigated area, with contour lines
  - Farm and field boundaries and water source or sources
  - Power points, such as electricity lines, in relation to water source and area to be irrigated, roads and other relevant general features such as obstacles
- data on water resources, quantity and quality over time, on water rights and on cost of water where applicable

- the climate of the area and its influence on the water requirements of the selected crops
- the soil characteristics and their compatibility with the crops and irrigation system proposed
- the types of crops intended to be grown and their compatibility with both the climate in the area, the water availability and the soils; current agricultural practices should be identified The next step is to analyse the farm data in order to determine the following preliminary design parameters:
- peak and total irrigation water requirements
- infiltration rate of soils to be irrigated
- maximum net depth of water application per irrigation
- irrigation frequency and cycle
- gross depth of water application
- preliminary system capacity

Principles of Design Adjustment

Once the preliminary design parameters are determined, the next phase is to reconcile them with the performance of the irrigation equipment and arrive at the final design. The final design steps involve:

- identification of irrigation system options with farmer participation
- preparation of system layout for the field shape and topography
- the hydraulic design and iterative adjustments
- irrigation equipment selection taking into consideration economic and financial aspects
- final irrigation system selection as well as options, taking into consideration farmers' preferences, management capabilities, labour aspects, financial capabilities and constraints.

The final design steps are intended to make the irrigation system selected compatible with the preliminary design factors. Each of the design steps is needed, irrespective of the irrigation system selected. However, the application of the final design steps varies between the periodic-move systems and the continuous-move systems. The differences are due to the fact that the periodic-move systems apply water for a set time while stationary before moving to the next position, while the continuous-move systems apply water while in motion. The final design process, for each of the periodicmove systems and continuous-move systems will be treated separately, in order to allow the specific aspects related to each system to come out clearly and be understood during the design process.

Steps for Preliminary Sprinkler Irrigation Design

The preliminary design factors that need to be established are: depth of water application per irrigation, irrigation frequency, duration of irrigation per set and required system capacity (flow rate). All these design parameters are derived from the data on climate, water, soil and plant.

Net Depth of Water Application

The depth of water application is the quantity of water, which should be applied during irrigation in order to replenish the water used by the crop during evapotranspiration. The computation of the net depth of water application requires the following inputs:

— the available soil moisture (FC-PWP)

— the allowable soil moisture depletion (P)

— the effective root zone depth of the crop (RZD)

Soil survey and tests should be done to determine the field capacity (FC) and permanent wilting point (PWP) of the soil. In the absence of equipment and time to do that, figures from literature, preferably local, can be used as estimates once the soil texture is known. However, published data on available moisture of different soil types do not always agree. Table 1 presents such data from two different sources.

Table 1. Available moisture for different major soil categories

From Israelson and Hansen		*From Withers and Vipond*	
Soil Category	*Available Moisture mm/m*	*Soil Category*	*Available Moisture mm/m*
Sandy	70-100	Sand	55
Sandy loam	90-150	Fine Sand	80
Loam	140-190	Sandy loam	120
Clay loam	170-220	Clay loam	150
Silty Clay	180-230	Clay	235
Clay	200-250		

The difference between field capacity and permanent wilting point will give the available soil moisture (water holding capacity), which is the total amount of water that the crop can use. Depending on the crop sensitivity to stress, the soil moisture should be allowed to be depleted only partially. For most field crops, a depletion of 50% of the available moisture is acceptable. This is the moisture that will be easily available to the crop without causing undue stress.

From past experience under irrigated conditions and similar climatic and soil conditions or from literature the effective root zone depth of the crop under consideration can be established. The maximum net depth to be applied per irrigation can be calculated, using the following equation:

$$d_{net} = (FC\text{-}PWP) \times RZD \times P \qquad (1)$$

Where:

d_{net} = readily available moisture or net depth of water application per irrigation for the selected crop (mm)

FC = soil moisture at field capacity (mm/m)

PWP = soil moisture at the permanent wilting point (mm/m)

RZD = the depth of soil that the roots exploit effectively (m)

P = the allowable portion of available moisture permitted for depletion by the crop before the next irrigation

In order to express the depth of water in terms of the volume, the area proposed for irrigation must be multiplied by the depth:

$$\text{Volume of water to be applied (m}^3\text{)} = 10 \times A \times d \quad (2)$$

Where:

A = area proposed for irrigation (ha)

d = depth of water application (mm)

Irrigation Frequency at Peak Demand and Irrigation Cycle

The peak daily water use is the peak daily water requirement of the crop determined by subtracting the rainfall (if any) from the peak daily crop water requirements.

Irrigation frequency is the time it takes the crop to deplete the soil moisture at a given soil moisture depletion level. After establishing the net depth of water application, the irrigation frequency at peak water demand should be determined using the following equation:

$$\text{Irrigation frequency (IF)} = \frac{d_{net}}{wu} \quad (3)$$

Where:

IF = irrigation frequency (days)

d_{net} = net depth of water application (mm)

wu = peak daily water use (mm/day)

Different crops require different amounts of water at the different stages of growth. From the meteorological data of the nearest meteorological station and using internationally recognised methods (e.g. Penman-Monteith) the crop and irrigation water requirements can be estimated. It should be mentioned that for design purposes we are particularly interested in the peak daily amount of water used by the crop, which is the worst case scenario.

The question arises as to whether the irrigation system should apply the d_{net} in 8, 7, 6, right down to 1 day. This choice will depend on the flexibility the farmer would like to have and his/her willingness to pay the additional cost for different levels of flexibility. If irrigation is to be completed in 1 day, the system becomes idle for the remaining 7 days, and the cost of the system would be exorbitant, since larger sizes of irrigation equipment would be required. On the other hand, for all practical purposes and in order to accommodate the time for cultural practices (spraying etc), it is advisable that irrigation is completed in less than the irrigation frequency. In the case of the example, 7 days irrigation and 1 day without irrigation is considered adequate. The 7 days required to complete one irrigation in the area under consideration is called the irrigation cycle.

Gross Depth of Water Application

The gross depth of water application (d_{gross}) equals the net depth of irrigation divided by the farm irrigation efficiency. It should be noted that farm irrigation efficiency includes possible losses of water from pipe leaks.

$$d_{gross} = \frac{d_{net}}{E} \tag{4}$$

Where:

E = the farm (or unit) irrigation efficiency.

Table 2. Farm irrigation efficiencies for sprinkler irrigation in different climates

Climate	*Farm Irrigation Efficiency*
Cool	80%
Moderate	75%
Hot	70%
Desert	65%

The farm irrigation efficiency of sprinkler systems varies from climate to climate. FAO proposed the above figures (Table 2).

Preliminary System Capacity

The next step is to estimate the system capacity. The system capacity (Q), can be calculated using Equation 5:

$$Q = \frac{10 \times A \times d_{gross}}{I \times N_s \times T} \tag{5}$$

Where:

Q = system capacity (m^3/hr)
A = design area (ha)
d = gross depth of water application (mm)
I = irrigation cycle (days)
N_s = number of shifts per day
T = irrigation time per shift (hr)

Final Design Steps for Periodic-Move Systems

Once the preliminary design parameters are obtained, the design adjustment can commence. The adjustment allows for the revision of the preliminary design parameters, in order to suit the physical, human, financial and equipment performance limitations or impositions. The next design step is to select the sprinkler and its spacing.

Sprinkler Selection and Spacing

The selection of the correct sprinkler depends on how the best fit spacing with a certain pressure and nozzle size can provide the water at an application rate that does neither cause runoff nor damage the crop and at the best possible uniformity under the prevailing wind conditions. The selected sprinkler should fully satisfy the irrigation water requirements and the irrigation frequency.

It is therefore necessary to know the infiltration rate of the soil before we can proceed with sprinkler selection. The infiltration rate can be determined using the double ring infiltrometers. In the absence of field data, the ranges of infiltration rate presented in Table 3 or any other literature can be used.

It should be pointed out that in order to avoid runoff the sprinkler application rate should not exceed the basic soil infiltration rate. Hence, the basic infiltration rate of the soil is used as a guide to select a sprinkler with a precipitation rate lower than the infiltration rate.

Table 3. Typical basic soil infiltration rates

Soil type	*Basic infiltration (mm/hr)*
Clay	1 - 7
Clay Loam	7 - 15
Silt Loam	15 - 25
Sandy Loam	25 - 40
Sand	>40

In this example, where a precipitation rate of 5-6 mm/hr is compatible with the soil and crop, there are several nozzle size, pressure and sprinkler spacing combinations to choose from, for example:

1. A 4.0 mm nozzle at 300 kPa and 12 m x 18 m spacing, gives a precipitation rate of 5.0 mm/hr
2. The same 4.0 mm nozzle at 350 kPa and 15 m x 15 m spacing, gives a precipitation rate of 5.16 mm/hr, and at 12 m x 18 m spacing, gives a precipitation rate of 5.37 mm/hr
3. A 5.0 mm nozzle at 300 kPa and 18 m x 18 m spacing, gives a precipitation rate of 5.25 mm/hr
4. The same 5.0 mm nozzle at the same spacing under 350 kPa, gives a precipitation of 5.68 mm/hr

Another aspect to consider in selecting a sprinkler is the energy cost. Lower pressures are preferable as long as the uniformity of application is not compromised. The Coefficient of Uniformity (CU) is a measure of the uniformity of water application. A value of 100% indicates perfect uniformity, which means that the water is applied to the same depth at each point in the field. As a rule, the selected sprinkler should have a CU of 85% or more. Where locally manufactured sprinklers are not tested for CU determination, it is advisable to avoid using the lowest pressure since usually this is the pressure that corresponds to low CU values.

Assuming that all three spacings fit the land, the next step is to find out how the winds will affect the spacing. For this purpose, the mean wind velocity of the windiest month of the year is considered. Most designers set the maximum spacing of sprinklers based on the information of Tables 4 and 5. It should be noted also that in the rectangular pattern better distribution is obtained when the lateral is placed across the prevailing wind direction. For variable wind directions, the square pattern gives better uniformity.

In the example, where the average wind velocity in September is 10 km/hr and in October 11 km/hr, the sprinkler spacing should be based on 50% of D for square pattern and 60% of D x 40% of D for rectangular pattern. The next step is to determine whether the three possible spacings above (15 m x 15 m, 12 m x 18 m, 18 m x 18 m) satisfy the wind requirements.

Table 4. Maximum sprinkler spacing as related to wind velocity, rectangular pattern

Average Wind Speed (km/hr)	*Spacing as Percent of Wetted Diameter (D)*
Up to 10	40% between sprinklers
	65% between laterals
10 - 15	40% between sprinkler
	60% between laterals
above 15	30% between sprinklers
	50% between laterals

From Table 5, the spacing for a square pattern for 11 km/hr wind speed is 15.25 m (0.5 x 30.50). Therefore, since 50% of D is greater than the 15 m spacing between sprinklers and 15 m spacing between the laterals, the wind requirement is satisfied. Similarly from Table 4, for a wind speed of 10-15 km/hr, 40% of D and 60% of D for the 12 m x 18 m spacing are 12.2 m (> than 12 m sprinkler spacing) and 18.3 m (> than 18 m lateral spacing) respectively. Therefore, the wind requirements are satisfied both for the 15 m x 15 m and the 12 m x 18 m spacing.

Table 5. Maximum sprinkler spacing as related to wind velocity, square pattern

Average Wind Speed (km/hr)	Spacing as Percent of Wetted Diameter (D)
Up to 5	55%
6 - 11	50%
13 - 19	45%

Let us determine whether the same sprinkler with a 4.0 mm nozzle would satisfy the wind requirements at the 12 m x 18 m spacing at 300 kPa. At this pressure, the wetted diameter is 26.60 m. 40% of D and 60% of D are 10.64 m (< than 12 m sprinkler spacing) and 15.96 m (< than 18 m lateral spacing) respectively. For the 15 m x 15 m spacing, 50% of D is 13.3 m (0.50 x 26.60), which is less than the sprinkler and lateral spacings of 15 m each. Therefor, the 4.0 mm nozzle operating at 300 kPa pressure does not meet the wind requirements either under 12 m x 18 m spacing or 15 m x 15 m spacing as the wetted diameter is too small compared to the desired spacing requirement.

Following the same procedure, the 5.0 mm nozzle does not meet the wind requirements at either 300 kPa or 350 kPa and 18 m x 18 m spacing. It meets the wind requirements at 12 m x 18 m and 15 m x 15 m spacing, but the precipitation rate of these last two spacings exceeds the soil infiltration rate. Therefore, they are not compatible with the infiltration rate of the soil. Hence, the 5.0 mm nozzle can not be considered. The 4.5 mm nozzle would meet the wind requirements at 300 kPa and a sprinkler spacing of 12 m x 18 m and 15 m x 15 m. However, the 12 m x 18 m spacing exceeds the infiltration rate of the soil.

In designing a sprinkler system the sprinkler precipitation rate should not exceed the infiltration rate of the soil. Moreover, a correction of the precipitation rate is recommended in order to avoid runoff in sloping land. Tables 6 and 7 are commonly used to assess the maximum precipitation rates under various conditions. In this case the slope of the land is 0.5%, therefore we do not need precipitation reduction.

Table 6. Maximum precipitation rates to use on level ground

Soil Type	Maximum Precipitation Rates (mm/hr)
Light sandy soils	18 - 12
Medium textured soils	12 - 6
Heavy textured soils	6 - 2.5

Table 7. Precipitation rates reduction on sloping ground

Slope Percent	*Reduction*
0 - 5%	0
6 - 8%	20
9 - 12%	40
13 - 20%	60
> 20%	75

Layout and Final Design

The system layout is obtained by matching the potentially acceptable spacings with the dimensions of the field such that as little land as possible is left out of the irrigated area. The layout should also accommodate access roads, drains and other structures such as toilets. The drains are not needed because of the irrigation method (like is the case in surface irrigation schemes), but to protect the scheme from high intensity rainstorms.

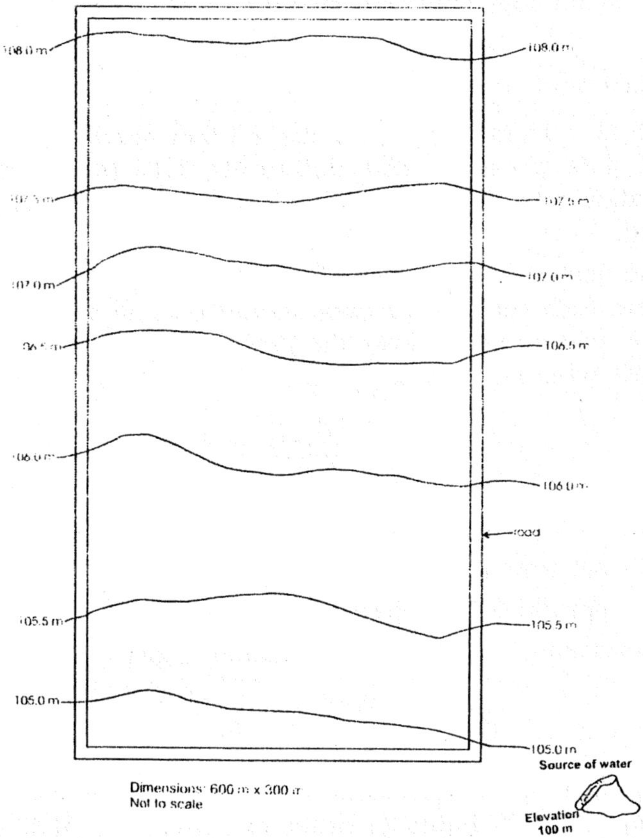

Figure 1. Farm map

The following sections illustrate the design procedure of the following periodic-move systems on the 18 ha (600 m length and 300 m width as shown in Figure 1) field example:

— Semi-portable sprinkler irrigation system for an individual farmer

— Semi-portable sprinkler irrigation system for a smallholder irrigation scheme

— Drag-hose sprinkler irrigation system for a smallholder irrigation scheme

Design of a semi-portable sprinkler irrigation system for an individual farm

The 15 m x 15 m spacing for the 4.0 mm nozzle operating at 350 kPa pressure and delivering 1.16 m^3/hr at an application rate of 5.16 mm/hr, was accepted as a potential spacing.

The next step is to determine the set time (T_s), which is the time each set of sprinklers should operate at the same position in order to deliver the gross irrigation depth, and establish whether it is acceptable.

$$T_s = \frac{d_{gross}}{P_r} \qquad (6)$$

Where:

T_s = set time (hr)

P_r = sprinkler precipitation rate (mm/hr)

Substituting the values in Equation 6 gives:

$$T_s = \frac{61.87}{5.16} = 11.99 \text{ hours}$$

Hence, each set of sprinklers should operate at the same position for 11.99 hours in order to deliver the 61.87 mm gross application per irrigation. If we assume that we are designing a permanent system this would have been ideal because we can have full utilisation of the equipment by having two sets per 24 hours. However, if we are designing a semi-portable system, where the laterals have to be moved from one position to the next, there would be no time available to move the laterals between each of the two shifts per day during the peak water demand period. In this case, we have the following choices:

1. To purchase twice the length of operating laterals so that extra laterals are moved while the other laterals are operating, or
2. To re-assess the moisture depletion level, or
3. To use a different sprinkler with the same or different spacing, nozzle, pressure and precipitation rate

As a rule, it is more economical to look into alternative 2 or 3 than to follow alternative 1. Alternative 2 involves readjusting the moisture depletion level. The effect will be a re-adjustment of d_{gross} and consequently the set time. During each irrigation we will apply the net equivalent depth to 7 days consumptive use. This would amount to a net application depth of 40.6 mm (7 x 5.8), which is equivalent to 41% (40.6/(140 x 0.7)) soil moisture depletion, with an irrigation frequency of 7 days. Allowing one day for cultural practices, the irrigation cycle would be 6 days. In order to apply the 40.6 mm net per irrigation, the gross application at 75% efficiency should be 54.1 mm (40.6/0.75).

Since the sprinkler precipitation rate is 5.16 mm/hr the sprinklers should operate for 10.5 hours (54.1/5.16) at each set during the peak demand period. With this adjustment more time is allowed (1.5 hours per set) to move the laterals. This makes the 15 m x 15 m acceptable in terms of the set time. It should be pointed out that the standard aluminium pipe lengths come in 9 metres and 6 metres. This means that each lateral will have to be composed of 9 metres and 6 metres pipes or 6 metres and 3 metres pipes. If for simplicity purposes the farmer would prefer to have the same length of pipes throughout each lateral, another spacing should be tested, say the 12 m x 18 m spacing.

The 12 m x 18 m spacing for the 4.0 mm nozzle operating at 350 kPa pressure and delivering 1.16 m^3/hr at 5.37 mm/hr precipitation rate can satisfy this requirement. This combination will be able to deliver the required 61.87 mm (with an 8 day frequency and 7 day cycle) in 11.5 hours or the re-adjusted application depth, 54.1 mm (with a 7 day frequency and 6 day cycle), in 10.0 hours. If the irrigation cycle of 7 days is adopted, then the time available for moving the pipes between each shift, is only 30 minutes. This option necessitates that labour for this purpose be available on Sundays. If the second alternative is adopted more flexibility is available as the system will operate for 6 days per week and a lot of time, 2 hours, is available for moving the pipes between each shift.

These changes will require an over-all re-assessment of the system capacity calculated earlier. It also remains to be seen how each of the two spacings fit the farm layout. After accepting spacings with promising set times, they should then be tried on the topographic map.

When preparing the layout of the system one should adhere to two principles, which are important for the uniformity of water application. Firstly, for the rectangular spacing the laterals should be placed across the prevailing wind direction. Secondly, where possible, laterals should run perpendicular to the predominant slope in order to achieve fairly uniform head losses.

Looking at the farm map, dimensioned at 300 m x 600 m net, and keeping in mind the above principles, it appears that some trade-offs should be made. With the main line placed at the 600 m long eastern boundary of the land, the 15 m x 15 m spacing would require 40 (600/15) lateral positions to cover the total area. Such a layout would permit the completion of irrigation in 5 days (40/(4 x 2)), with 4 laterals operating at a time and 2 shifts per lateral per day. Therefore, the 15 m x 15 m spacing with 4.0 m m nozzles operating at 350 kPa and applying 5.16 mm/hr would deliver the amount of water required for 7 day frequency (7 x 5.8/0.75 = 54.1 mm) in 5 days, with an operation of 2 shifts per day for 10.5 hrs per shift. The 15 m x 15 m spacing would therefore be a suitable spacing, but from the point of view of the utilisation of the invested capital the system would only be utilised for 71% (5 days out of a possible 7 days) of the time, at peak demand. The capacity of such a system can be calculated using Equation 7:

$$Q = Nc \times Ns \times Qs \quad (7)$$

Where:

Q = system capacity (m^3/hr)

Nc = the number of laterals operating per shift

N_s = the number of sprinklers per lateral

Q_s = the sprinkler discharge.

N_s is obtained by dividing the length of the lateral by the sprinkler spacing. In this case, the lateral is 300 m and the sprinkler spacing is 15 m, therefore Ns is 20.

Substituting N_c, N_s and Q_s in the equation gives:

$$Q = 4 \times 20 \times 1.16 = 92.8 \text{ m}^3/\text{hr}.$$

In comparison to the optimum theoretical capacity (preliminary system capacity) calculated earlier at 72.3 m^3/hr, for a 7 day cycle, this flow (92.8 m^3/hr) is higher. Higher flows than necessary imply bigger pumping units and larger diameters of pipes, which increase the capital investment required for the system.

Alternatively, if 3 laterals are used per shift and there are 2 shifts per day, the irrigation cycle can be completed in 6.5 (39/(3 x 2)) days, covering 39 positions. Consequently, a strip of land of 0.45 ha (15 m x 300 m) will not be irrigated. This is a more economical approach reducing the number of laterals to 3, and the system capacity to 69.6 m^3/hr (20 x 3 x 1.16). However, this requires labour every day, including Sundays, which is a disadvantage of such a layout. There is a need to compare the 15 m x 15 m spacing with the other potential spacings such as the 12 m x 18 m.

While the main line is maintained at the eastern boundary of the land, let us try to see how the 12 m x 18 m spacing fits. Within the 600 metres length of field, 33.3 (600/18) lateral positions, with 25 (300/12) sprinklers each can fit. If we operate 3 laterals at a time for 2 shifts per day, 33 positions will be covered in 5.5 days (33/(3 x 2)), leaving 0.3

positions not irrigated. If one lateral is used alone for this position then the pressure at the sprinkler nozzle and consequently the discharge will be too high for a uniform water application. Therefore, in order to use this spacing a strip of land of 6 m wide and 300 m long would have to be excluded from irrigation, resulting in a reduction of the area by 0.18 ha. Such a layout would result in a system capacity of 87 m^3 /hr (3 x 25 x 1.16).

Therefore, the farmer and the designer would have to choose between: (a) a layout of 15 m x 15 m that can cover the whole area at a higher cost per unit area and also with laterals composed of 6 m and 3 m lengths of pipes, (b) a layout of 15 m x 15 m that reduces the cost but also reduces the area by 0.45 ha and (c) a layout based on 12 m x 18 m with moderate cost, convenience in operation (laterals are composed of 6 m lengths) and able to complete irrigation in 5.5 days per week but reducing the area by 0.18 ha.

Assuming that for practical and economic reasons the farmer agreed to adopt the 12 m x 18 m spacing and lose a small strip of land, he/she is still faced with another practical problem with both approaches. After all, when the laterals reach the last position they have to be returned to the first position. This would require the transport of 900 (300 x 3) metres of pipes for a distance of 198 (11 x 18) metres in the case of the 12 m x 18 m spacing, or 1200 (300 x 4) metres of pipes for a distance of 150 (10 x 15) metres in the case of 15 m x 15 m spacing. A more favourable arrangement from the operational point of view can be attained by locating the main line in the middle of the plot and in parallel to the length of the field. Such a layout will permit the rotation of the laterals around the mainline.

In this case the adopted spacing (12 m x 18 m) would require 3 short laterals with 13 sprinklers each and another 3 short laterals with 12 sprinklers each operating at the same time in order to complete the 66 lateral positions in 5.5 days {66 positions/(6 laterals x 2 shifts/day)} with the same flow (Q = 87.0 m^3/hr).

Basically, the choice will depend on the economics. One can compare costs to establish whether the savings obtained using smaller diameter laterals can compensate, and to what extent, for the additional cost of main pipe to transport the water from the edge of the farm to the middle point of the southern border.

Allowable pressure variation

Pressure differences throughout the system or block or subunit should be maintained in such a range so that a high degree of uniformity of water application is achieved. Addink *et al* and Keller suggest that for practical purposes the allowable pressure loss due to friction can be estimated at 23.4% of the required average pressure. For the same reason, the friction losses in the lateral should be kept to a minimum. Other sources suggest that allowable pressure variation should not exceed 20% of the sprinkler operating pressure. Of the 12 m x 18 m spacing for the 4.0 mm nozzle operating at 350 kPa, the allowable pressure variation in the system should not exceed 20% of the sprinkler operating pressure, which is 70 kPa (350 x 0.2) or 7 metres.

Pipe size determination

Pipe size determination involves selecting the diameter of a pipe type, which can carry a given flow at or below the recommended velocity limit. For example, the velocity limit for uPVC pipes is about 2 m/s. Also, depending on the water pressure, different classes of pipes can be selected for the same pipe type. uPVC pipes come in pressure ratings of 40 metres (Class 4), 60 metres (Class 6), 100 metres (Class 10) and 160 metres (Class 16). If, for example, the water pressure at a pipe section is 30 metres and uPVC pipe is being used, then a pipe rated at class 4 should be selected. There are a number of different types of pipes.

Allowable pressure variation

Pressure differences throughout the system or block or subunit should be maintained in such a range so that a high degree of uniformity of water application is achieved. Addink *et al* and Keller suggest that for practical purposes the allowable pressure loss due to friction can be estimated at 23.4% of the required average pressure. For the same reason, the friction losses in the lateral should be kept to a minimum. Other sources suggest that allowable pressure variation should not exceed 20% of the sprinkler operating pressure. The example, of the 12 m x 18 m spacing for the 4.0 mm nozzle operating at 350 kPa, the allowable pressure variation in the system should not exceed 20% of the sprinkler operating pressure, which is 70 kPa (350 x 0.2) or 7 metres.

Pipe size determination

Pipe size determination involves selecting the diameter of a pipe type, which can carry a given flow at or below the recommended velocity limit. For example, the velocity limit for uPVC pipes is about 2 m/s. Also, depending on the water pressure, different classes of pipes can be selected for the same pipe type. uPVC pipes come in pressure ratings of 40 metres (Class 4), 60 metres (Class 6), 100 metres (Class 10) and 160 metres (Class 16). If, for example, the water pressure at a pipe section is 30 metres and uPVC pipe is being used, then a pipe rated at class 4 should be selected. There are a number of different types of pipes. The engineer should consider what pipes are available on the market and their costs.

Laterals

Laterals in a semi-portable system are aluminium pipes with multi-outlets (sprinklers) along their length. The friction losses, either calculated or obtained from charts, have to be corrected since the flow reduces along the lateral. This is done by using Christiansens adjustment factor "F". Table 8 shows Christiansens F values for velocity exponent m = 2.0, most commonly used in sprinkler irrigation systems. Alternatively, the friction losses in every segment of the pipe can be calculated using the relevant charts for corresponding flow for each section.

Table 8. Christiansens "F" factors for various numbers of outlets

Number of outlets	*F for m = 2.0*	*Number of outlets*	*F for m = 2.0*
1	1.000	14	0.370
2	0.625	15	0.367
3	0.518	16	0.365
4	0.469	17	0.363
5	0.440	18	0.361
6	0.421	19	0.360
7	0.408	20	0.359
8	0.398	21	0.357
9	0.391	22	0.355
10	0.385	23	0.353
11	0.380	24	0.351
12	0.376	25	0.350
13	0.373		

Main line

It is necessary to know some characteristics of some of the pipes commonly used in irrigation, unplasticised Polyvinylchloride (uPVC) pipes and Asbestos-cement (AC) pipes. AC pipes are no longer recommended for use in domestic water supply systems, because of the health hazard to workers in the manufacturing and installation of these pipes as well as the public at large. Therefore, especially where the main supply line is integrated with a domestic water supply, AC pipes should not be used. The pressure within any part of the pipe network should not exceed the working pressure of that pipe, in order to comply with established standards. This should be kept in mind when selecting pipe sizes for frictional loss calculations. In addition, the recommended maximum velocities (2 m/s) should not be exceeded.

AC pipes normally come in 4 metres lengths. Seven different classes of asbestos cement pipes are usually manufactured (Table 9). The most common sizes are 50-900 mm nominal diameter, even though larger sizes can be manufactured. Each pipe length is marked with the size and class of the pipe.

Table 9. Asbestos-cement pipe classes and corresponding pressure rating

Class	*Working Pressure (kPa)*	*Test Pressure (kPa)*
6	300	600
12	600	1 200
18	900	1 800
24	1 200	2 400
30	1 500	3 000
36	1 800	4 200
42	2 100	4 200

While the class 6 pipe is used for surface irrigation the most commonly used classes for pressurised irrigation systems are the classes 12, 18 and 24. All AC pipes and fittings are only recommended for underground installation, as they can easily be damaged or dislocated by, for example traffic, agricultural implements and animals. uPVC pipes normally come in 6 metres lengths. The most commonly available uPVC pipes fall in 4 to 16 pressure classes shown in Table 10.

Table 10. uPVC pipe classes and corresponding working pressure rating

Class	*Working Pressure (kPa)*
4	400
6	600
10	1 000
16	1 600

The most common sizes range from 25 mm to 250 mm in diameter. In the example, the position of each lateral affects the friction losses in the main line since it affects the flow at the different sections of the main line. Therefore, friction losses corresponding to different alternative positions of the laterals should be analysed. As a rule, the highest friction losses in the main occur when all laterals operate in the middle position, which is position 6 in the example. This rule does not hold true always. Therefore, the friction losses in the main are calculated for the first, middle and last positions. Of the three calculations, the highest is used for the compilation of the total head losses and the selection of the relevant pipe class.

Using the frictional loss chart for uPVC pipes, the friction losses of the main line can be calculated as shown below.

Q = the discharge or flow rate within that section of the pipe, the units depending on the chart being used (in this case m^3/hr)

L = the length of pipe for that section (m)

D = the pipe size diameter (mm)

HL = the friction loss of the pipe (m)

Pipe class shows the working pressure of the pipe, not to be exceeded in that section.

The frictional loss charts also show the recommended maximum velocities of flow in the pipes. The smaller the velocity, the less the head loss per unit length of pipe. The higher the flow, the higher the friction loss per unit length and the more it is turbulent. This leads to the possibility of higher internal wear of the pipe and possibility of water hammer, when the system is shut down suddenly.

Total head requirements

The total head requirements are composed of the pump suction lift, the friction losses in the supply line, the friction losses in the main, lateral and fittings, the riser, the sprinkler operating pressure and the difference in elevation.

The suction lift is the difference in elevation between the water level and the eye of the pump impeller plus the head losses in the suction pipe. The head losses of the suction pipe comprise the frictional losses of the pipe, fittings and the velocity head. The friction losses of the suction pipe are insignificant compared to the velocity head, if the pipe is short.

$$\text{The velocity head is equal to } \frac{V^2}{2g}$$

Where:

v = water velocity (m/s)

g = acceleration due to gravity (9.81 m/s^2)

Keller and Bliesner recommend that for centrifugal pumps the diameter of the suction pipe should be selected such that the water velocity $v<3.3$ m/s in order to ensure good pump performance. Assuming this maximum velocity for the flow and applying the above formula, then the velocity head corresponding to the minimum diameter of the suction pipe that can be selected to satisfy this condition is 0.56 m (3.3^2/ (2 x 9.81)).

In the example, assuming that the water level is at 99 m, the difference in elevation between the water and the eye of the impeller (located at 100 m) is 1 m. Since the maximum velocity head is 0.56 m, the suction lift is 1.56 m (1 + 0.56). Assuming minor losses in fittings and a short suction pipe, the suction lift is rounded up to 2 m.

The difference in elevation is the difference between the ground level of the sprinkler, located at the highest point, and the eye of the pump impeller. This is obtainable from the contour map and is approximately 8 m (108 - 100).

The length of the supply line is 251 m (70 m from pumping station to field edge plus 150 m from field edge to the middle of the field, plus 4 m for the road, plus 27 m to the first hydrant. The friction losses for the supply line are computed as follows:

$$Q = 87 \text{ m}^3/\text{hr}$$

$$L = 251 \text{ m}$$

$$D = 200 \text{ mm}$$

$$HL = 0.0035 \times 251 = 0.88 \text{ m}$$

For friction losses in the riser we can assume about 0.25 m per m of riser and for fittings we usually take 10% of the total head losses. Additionally, the height of the riser should be included in the calculations. This is assumed to be 2 metres in the example, to cater for tall crops such as maize. In the example, the total head requirements would be as shown in Table 11.

Table 11. Total dynamic head requirements for a semi-portable system for an individual farm of 18 ha

Total Dynamic Head Component	*Head Loss (m)*
Suction lift	2.00
Supply line	0.88
Main line	2.49
Lateral	0.96
Riser	2.50
Sprinkler Operating Pressure (SOP)	35.00
Subtotal	43.83
Fittings 10%	4.38
Elevation difference	8.00
Total	56.21

Pump selection and power requirements

The next item to be dealt with in a design is the selection of the pump and power plant. From the manufacturers charts, a pump which should provide the desired head and flow at the highest possible efficiency and an electric motor to drive the pump should be selected. Also of great importance in the selection of a pump is that the Net Positive Suction Head Available (NPSHA) exceeds the Net Positive Suction head Required (NPSHR) by the pump. The basic formula for power requirement (kW or BHP), calculations is provided below:

$$\text{Power requirement in BHP} = \frac{Q \times TDH}{273 \times E_p} \tag{8}$$

or

$$\text{Power requirement in kW} = \frac{Q \times TDH}{360 \times E_p} \qquad (9)$$

Where:

kW or BHP = energy transferred from the pump to the water

Q = discharge (m^3/hr)

TDH = total dynamic head (m)

E_p = the pump efficiency (%) from the pump performance chart

360 & 273 = conversion constants for metric units.

It should be noted that this formula is an expression of the actual power required at the pump.

For the example:

$$\text{Power requirements in BHP} = \frac{87 \times 56.21}{273 \times 0.6} = 29.86$$

or

$$\text{Power requirements in kW} = \frac{87 \times 56.21}{360 \times 0.6} = 22.64$$

Depending on the losses in transferring the power to the pump, an allowance of 20% should be made, thus an engine of 35.83 HP (2.01 HP/ha) or 27.17 kW (1.53 kW/ha) should be ordered. Depending on the manufacturer and market availability, the next larger size of motor or engine should be obtained.

Design of a semi-portable sprinkler irrigation system for a smallholder scheme

Smallholder irrigation schemes are here defined as irrigation schemes of any size in which individual farmers have small plot holdings. Most of the irrigation infrastructure, such as pumping units, mainlines, secondary lines and tertiaries is communally operated. Depending on the design, even laterals and sprinklers could be communally operated. In some cases semi-portable sprinkler systems designed for smallholder irrigators were designed as if they were to serve commercial farms and the land was sub-divided into small holdings and allocated to communal farmers. However, because of the communal use of laterals among many plot holders and the resultant lack of accountability for damages, replacement of laterals every 4-5 years is not uncommon. Under careful handling aluminium laterals have a life expectancy of 15 years. Also since smallholders as a rule grow 2-4 crops per season these systems were not providing for optimum irrigation as they were designed for monoculture.

The approach described below is based on the principle that each plot holder will have his/her own in-field irrigation equipment (laterals, sprinklers and risers) to serve 3-4 crops per sea-son. The rest of the equipment remains communally operated. Countries that have such schemes include Zimbabwe, Kenya, Swaziland and South Africa.

Sprinkler selection and spacing

Referring to Figure 1, the 18 ha plot (300 m x 600 m) can be sub-divided into 32 plots of approximately 0.5 ha each, allowing for road access to each plot from all sides. If plots of 70 m x 72 m are demarcated, then a strip of land of 12 m x 300 m will remain unutilised, when 4 metres wide strips, consisting of in-field roads and surface drains, are incorporated into the layout. Alternatively, 72 m x 72 m plots can be demarcated and the area is extended by a strip of 4 m x 300 m. Assuming that this is possible the second option is favoured (Figure 2). The best-fit sprinkler spacing for a plot size of 72 m x 72 m would be 12 m x 12 m. This plot size and sprinkler spacing results in 12 lateral positions, if the tertiary line dissects the plot.

The 3.0 mm nozzle at various pressures would more than satisfy the wind requirements criteria, the spacing being within 50% of the diameter of coverage (0.5 x 25m = 12.5 m). The same applies to the 3.5 mm nozzle sprinkler and the 4.0 mm nozzle sprinkler. However, the precipitation rate of the

4.0 mm nozzle exceeds the 5-6 mm/hr soil infiltration rate limit in the example. The same holds true for the 3.5 mm nozzle when it operates at more than 300 kPa.

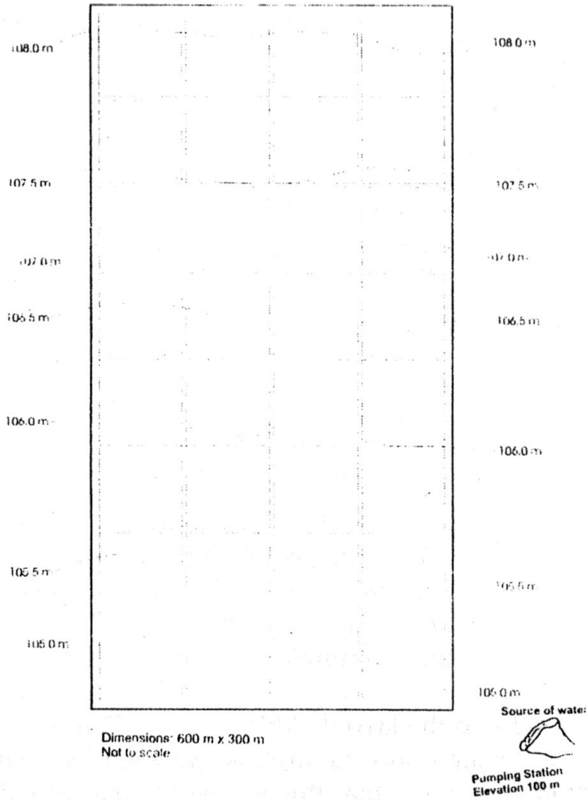

Figure 2. Plot layout in smallholder scheme

Also, as a rule, the water distribution from a sprinkler is not as good when it operates at the extreme low end of its operating pressure range. Hence, both the 3.0 mm and 3.5 mm nozzles will not be considered at 250 kPa. Of the remaining combinations, the best fit from the operational point of view and capital cost as well as energy requirements will be considered.

Using the basic farm data given in Example 1, the following sprinkler options are considered:

1) 3 mm nozzle, 300 kPa, 4.38 mm/hr at 12 m x 12 m spacing
   8 day frequency, 7 day cycle,
   d_{gross} = 61.87 mm at 47% depletion.

$$\text{Hours of operation per shift} = \frac{61.87}{4.38} = 14.12$$

This option allows for only one shift per day of 24 hours. As there are 12 lateral positions per plot, the irrigation cycle cannot be completed in 7 days unless 2 laterals are provided per plot. This is a possibility, but would be a costly option.

2) 3 mm nozzle, 300 kPa, 4.38 mm/hr at 12 m x 12 m spacing
   7 day frequency, 6 day cycle,
   d_{gross} = 54.13 mm at 41% depletion.

$$\text{Hours of operation per shift} = \frac{54.13}{4.38} = 12.35$$

This option has the same drawback as the previous alternative.

3) 3 mm nozzle, 350 kPa, 4.72 mm/hr at 12 m x 12 m spacing
   8 day frequency, 7 day cycle,
   d_{gross} = 61.87 mm at 47% depletion.

$$\text{Hours of operation per shift} = \frac{61.87}{4.72} = 13.10$$

The same drawback also prevails here.

4) 3 mm nozzle, 350 kPa, 4.72 mm/hr at 12 m x 12 m spacing
   7 day frequency, 6 day cycle,
   d_{gross} = 54.13 mm at 41% depletion.

$$\text{Hours of operation per shift} = \frac{54.13}{4.72} = 11.46$$

This alternative appears to be promising. It allows for two shifts per day (24-hour period) with half an hour available between each shift to move the portable lateral to the next operating position. It also allows a high degree of equipment utilization, as the system will operate for about 23 hours per day during the peak demand period. While the other options would have necessitated providing two laterals per plot, this option will only require one lateral and therefore saves on costs. This option is therefore adopted.

System layout

When preparing the layout of the pipe network, a degree of flexibility in operation should be considered. At the same time, the designer has to adhere to the criteria for locating the portable lateral in relation to the wind direction and the land slope, wherever possible. One option would be to provide for a main line in the middle of the land running against the main slope. The secondary pipelines would then run parallel to the contours and serve pairs of plots on each side. Tertiary pipelines would take off from the secondary line and supply the lateral of each plot as shown in Figure 3.

While the first option is more economical, the second option is more flexible since any breakages of any tertiary pipeline would affect the operation of only one plot compared to two plots in the first option. Naturally it may be argued that more economical layouts can be derived by having more than two plots served by one tertiary line. In contrast to that, individual water control may be necessary for schemes with a big number of plot holders as it allows for better water management. These options need to be discussed with the farmers so that they can select which one they prefer. Therefore, the following design will be based on the layout of Figure 3.

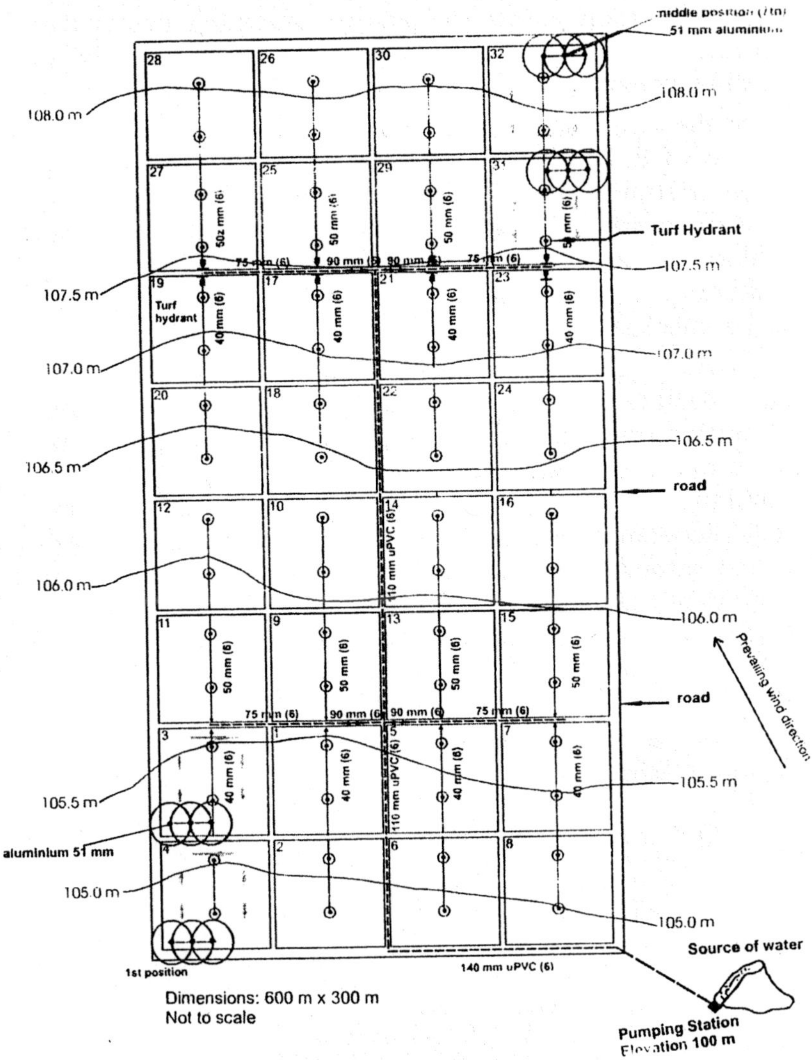

Figure 3. Semi-portable system layout based on 12 m × 12 m spacing with tertiaries serving two plots (first attempt for pipe sizing)

Pipe size determination

The pipes should be sized in such a way that the total pressure variation within the system (assuming this is the hydraulic units) does not exceed 20% of the sprinkler operating pressure.

For the case under consideration it should not exceed 70 kPa (350 x 0.2) or 7 metres. This means that the head losses due to friction and difference in elevation between the reference point (beginning of first secondary) and the sprinkler of the furthest plot with the highest elevation should not exceed 7 metres. The hydraulics of each plot must also be checked and conform to this principle.

A point of clarification may be required at this stage. The total allowable pressure variation is applied to a hydraulic unit. Such a unit may be the total area of the scheme or any part of that area. Hence, a reference point may be the first secondary offtake or every secondary or every tertiary offtake depending on how steep the land is. In the example, the first secondary offtake located at elevation 105.6 metres is considered as the reference point, with the whole scheme considered as one hydraulic unit.

Laterals

The following friction losses are calculated for the 30 m long aluminium lateral:

$$Q = 3 \text{ sprinklers} \times 0.68 \text{ m}^3/\text{hr per sprinkler}$$
$$= 2.04 \text{ m}^3/\text{hr}$$
$$L = 30 \text{ m } (12 \text{ m} + 12 \text{ m} + 6 \text{ m})$$
$$D = 51 \text{ mm}$$
$$HL = 0.0023 \times 30 \times 0.518 = 0.036 \text{ m}$$

Naturally, for such a small discharge a smaller diameter pipe could be used, say the 32 mm, if available. In this case the friction losses are:

$$HL = 0.018 \times 30 \times 0.518 = 0.28 \text{ m}$$

Since in most countries of East and Southern Africa the 32 mm aluminium pipe is not available, we will maintain the 51 mm diameter lateral in the example. As each valve hydrant serves 3 lateral positions (one on each side of the hydrant plus one at the hydrant), the friction losses for a 12 m aluminium pipe (header) with a flow of 2.04 m^3/hr should be included:

$$HL = 0.0023 \times 12 = 0.028 \text{ m}$$

Therefore, the total friction loss of the lateral when the header is used is 0.064 m (0.036 + 0.028).

Tertiaries

The tertiary lines should be buried so that they do not interfere with cultural practices. As such, they should preferably be made of non-corrosive material. This limits the choice to either uPVC or AC. As both types have similar hydraulic characteristics, the criteria for choice would be availability, cost, health aspects and ease of installation. As a rule, uPVC is more compatible to the above criteria for small sizes and AC for large sizes. Again, it varies from country to country. In the case under consideration, uPVC has been adopted.

As the flow in some of the tertiaries of this example is directed uphill and in others downhill, friction losses for two cases will be calculated and later on checked against the total allowable pressure variation that the hydraulics of each plot should conform to. For each case the hydraulics of the tertiaries are calculated separately for lateral position 1 and lateral position 7 in order to establish the worst situation.

Referring to the allowable pressure variation of 7 m and taking into consideration the head losses in the tertiaries and laterals of 3.06 m (0.064 + 3.00) at position 1, a balance of 3.94 m (7 - 3.06) is left for the secondary, mainline and the difference in elevation. However, since there is a negative difference in elevation of 1.1 m (104.5 - 105.6), the total available head for the secondary is 5.04 m (3.94 + 1.1). In position 7, the negative difference of elevation is 0.4 m (105.2 - 105.6). Following the same logic, the total head available for the secondary will be 5.64 m {7 - (0.064 + 1.70) + 0.4}. Since the secondary must satisfy the needs of both position 1 and position 7 and the head losses plus the difference in elevation do not exceed the allowable pressure variation, the lowest available head (5.04 m) is the one selected. The secondaries would thus be sized so that the head losses do not exceed 5.04 metres.

Since the total allowable pressure variation is 7 metres and about 1.04 m (0.064 + 0.98) is lost to friction in the laterals and uphill flow tertiaries, 5.96 m (7 - 1.04), is available to cover the 2.65 m (108.25 - 105.6) dif-ference in elevation between the first set of secondary take-offs from the main and the highest plot as well as the friction losses in the secondaries and the main line.

Secondaries

All secondary lines can be of the same size, since they carry the same flow, and they run parallel to the contours. However, because of the 4 metres wide road and the position of the mainline at the east side of the road, the eastern secondaries are 36 metres long each and the western secondaries are 40 metres (36 + 4).

The head losses in the first part of the eastern secondaries are given by:

The second sections of the secondaries are all the same length at 76 m (36 + 4 + 36) and their head losses are calculated as follows:

$Q = 8.16\ m^3/hr$

$L = 76$ m (36 + 4 + 36)

$D = 75$ mm uPVC (6)

$HL = 0.0054 \times 76 = 0.41$ m

Thus, HL (total) for the secondary line is 0.70 m (eastern side) or 0.73 m (western side)

Mainline

For purposes of allowable pressure variation computations, the main is considered to start at the take-off point of the first two secondary lines. Therefore, the friction losses are obtained as follows:

Q = 32.64 m^3/hr (16 x 2.04) (plots 17-32)

L = 304 m (4 x 72 + 4 x 4)

D = 110 mm (6)

HL = 0.0098 x 304 = 2.98 m

Thus HL for the mainline is 2.98 m

While it appears that the total available head of 5.96 m for the difference in elevation and the head losses in the main and secondary lines is not exceeded, it is necessary to carry out verification on a plot by plot basis.

Hydraulics of individual plots in relation to the total allowable pressure variation

As pointed out earlier, the total head required by each plot in terms of friction losses and elevation differ-ences from the reference point should not exceed 7 m. Hence, if for example the plot with the lowest friction losses and elevation difference from the reference point is -1 (minus 1)m then the plot with the highest should not exceed 6 m. Also, a change in the diameter of the main from 110 mm to 140 mm will reduce the friction losses by 2.01 m (2.98 - 0.97) for the plots where the allowable pressure variation exceeds the 7 m limit, hence:

Q = 32.64 m^3/hr

L = 304 m

D = 140 mm PVC (6)

HL = 0.0032 x 304 = 0.97 m

By changing the size of the secondaries serving plots 1-16 from 90 mm and 75 mm to 63 mm, a saving in the cost can be made while the frictional losses are brought closer to those of plots 17-32, bearing in mind that the pump will have to satisfy the worst case anywhere. The friction losses for plots 1-16 are re-calculated as follows:

Q = 16.32 m^3/hr (8 x 2.04)

L (eastern side) = 36 m

D = 63 mm uPVC (6)

HL (eastern side) = 0.04 x 36 = 1.44 m

The head losses in the first part of the western secondaries are given by

Q = 16.32 m^3/hr (8 x 2.04)

L (western side) = 40 m

D = 63 mm uPVC (6)

HL (western side)= 0.04 x 40 = 1.60 m

The second sections of the secondaries are all the same length at 76 (36 + 4 + 36) m and their head losses are calculated as follows:

Q = 8.16 m^3/hr

L = 76 m

D = 63 mm uPVC (6)

HL = 0.0124 x 76 = 0.94 m

Thus, HL (total) for the secondary line is 2.38 m (eastern side) or 2.54 m (western side)

Changing the size of the uphill tertiaries of plots 9-16 to 40 mm results in cost savings while the losses are maintained within the allowable pressure variation.

References

Addink, *et al.*, *Design and operation of farm irrigation systems*, ASAE, 1989.

Kay, M., *Sprinkler Irrigation: Systems and Practice*, Bedford, 1986.

Keller, J. and Bliesner, R.D., *Sprinkler and Trickle Irrigation*, Chapman and Hall, New York, 1990.

Keller, J., *Sprinkler and Trickle Irrigation*, Utah State University, Utah, 1989.

3

Sprinkler Irrigation Design

Sprinklers require two things to operate-water flow and water pressure. When a sprinkler system design fails, it is almost always the result of a lack of water pressure. Go get a sprinkler, any sprinkler will do fine, even one you put on the end of a hose. Now get a big bucket of water. 5 gallons is an ideal size. Now place the sprinkler on the ground (somewhere outside is strongly recommended) and pour the water into the sprinkler inlet. The water you poured into it didn't have any pressure to make the sprinkler operate.

Water pressure is the energy that makes the sprinkler do its thing. Water pressure is like the gasoline for the car. Flow is like the distance the car travels. No gas, no travel. No water pressure, no water flow.

Now think of the inverse of the above example, because we're going to be using it in our sprinkler design. We know that water requires pressure to flow. It also stands to reason that just as the car uses up the gas when it travels, the water also uses up the pressure as it flows! So from the moment the water enters your yard, it is losing pressure. It loses some pressure when it goes through a water meter. It loses some more pressure as it moves through the pipes. Squeezing through the backflow preventer and valves eats up more pressure. But here is the key- after it gets to the sprinkler head there still must be enough pressure left to make the sprinkler head work.

Sprinkler design is simply manipulating out how much water pressure is lost between the point the water enters your yard and the sprinkler head.

Measure Your Yard

Gathering information is the first step for most projects and it is one of the most important steps when designing both sprinkler and drip irrigation systems. A mistake at this point in the process will affect everything else, so accuracy and care are important. Although the references here are to a residential yard, the principles apply equally well to other areas. Here are a few tips for getting started.

1. *Use a measuring tape or other accurate measuring device.* Many irrigation system problems are traced back to inaccurate measurement of the area to be irrigated. Go ahead and round off numbers to the nearest foot. But keep in mind that a minor deviation of a couple of feet can result in dry spots!
2. *Draw the area to be irrigated to scale on a piece of graph paper as you measure it.* Use a scale of 1 square = 1 foot, or 1 square = 5 feet if needed. Drawing as you go allows you to quickly catch errors. For instance, say you measure the perimeter of the yard. Eventually you get back to the point where you started, but you find that the lines drawn on your paper don't meet like they should! Now you know there is an error in one of the measurements, and you can go back and find it quickly, while it's all still fresh in your mind.
3. *If there is a building adjoining the irrigated area, start by measuring the building's perimeter.* Most buildings are built with square corners and straight sides, which gives you a good, accurate base for the rest of your measurements.
4. *Use the straight sides of the building as a guide for establishing the perimeter of the area to be irrigated.* Measure in a straight line the distance from each corner of the building to the edge of the irrigated area or property.

Use the straight edge of the building to line up your measurements. Now you can draw in the property lines on your graph paper using these measurements. Double check everything by measuring the length of each property line. It should match exactly what you have drawn. Don't be surprised if you find that an area you thought was "rectangular" is actually lopsided like the one in the sample drawing! Most pieces of property aren't perfect rectangles.

5. The final step is to draw in the other items that are present in the area. For example, if this is your house, you would need to show the edges of lawns, tree locations, the location of your pool house, the servant's quarters, etc. With the building and property lines drawn it is easy to use the grid on your graph paper to establish the locations of these items.

Selecting Your Equipment

Collect Pressure Loss Data

When selecting your sprinkler equipment we need to consider pressure loss in the irrigation system. Like all other mechanical systems an irrigation system consumes energy when it operates. The irrigation system uses energy in the form of water pressure which, as we noted earlier, we will be measuring in PSI (pounds per square inch). Each component in the irrigation system that the water passes through uses up a little bit of that water pressure. If we run out of water pressure before the water makes it through the system, then the irrigation system will not work (no water pressure). So each loss of water pressure must be calculated and adjusted if necessary to assure that the irrigation system will work properly. This is done by means of a trial and error process.

Pressure Regulators and Pressure Reducers

A pressure reducer is another name used for pressure regulators. It is not commonly used in the industry, so I'm going to use the preferred name "pressure regulator". If you have a pressure regulator then you get to take a little shortcut. On your pressure loss table you get to ignore the pressure loss for everything upstream of the pressure regulator. But before you decide to really take a shortcut and install the pressure regulator right before the valves, consider that the higher pressure may not be good for those upstream components. Don't forget the pressure setting of a pressure regulator must always be at least 15 PSI lower than the incoming pressure. So if the incoming pressure is 80 PSI the pressure regulator must be set at 65 PSI or less. Otherwise the pressure regulator will not work accurately and will allow damaging pressure surges to pass through.

Water Meters

The water meter isn't usually something you get to select. You either have one, or you don't! If you buy your water from a water company or municipal water district you probably have a meter, if not you probably don't. The water meter measures how much water you use, so the water company can charge you for it. If you do have a meter it will most likely be located near where your water supply taps into the water company's pipes. This is usually right next to the curb, or on the side of the alley. In some cold weather areas the meter is inside the house basement. If the meter is buried in a box, be careful when you open the box. The meter should have a size stamped on it, if not, call your water company and ask them for the meter size, it should be in their water billing information.

Backflow Preventers

All irrigation systems contain a chemical called dihydrogen monoxide. This odourless, tasteless chemical is known to be deadly if inhaled and has killed millions of people. OK, before you flame me, dihydrogen monoxide is, of course, the scientific name for ordinary water (H_2O). Which points out that you have to be really careful about what you read and believe on the Internet. Unfortunately when it comes to backflow preventers there is a lot of questionable, and sometimes completely wrong, information. So what's the deal with backflow prevention? Is it really necessary? The answer is yes!

Your landscape has all kinds of nasty things in it that will make you sick or worse if you drink them. Thus irrigation water is considered a contaminant (creates a health hazard) rather than just a pollutant (is objectionable in colour or odour). What's in irrigation water? How about toxic chemicals (fertilisers, pesticides, etc.) and animal waste? These things can and will come back up your irrigation pipes and into your drinking water if you don't stop them. If you have a well they can go down your well and into everyone else's drinking water. If you are on a community water system they could go back up into the pipes and poison your neighbours. The valves that turn on and off your irrigation system are not sufficient to stop backflow. The purpose of the backflow preventer is to protect you when the valve breaks or leaks, which all valves will do eventually. Saving a little money by skipping the backflow preventer will not seem so smart after you spend a small fortune on hospital bills (or funeral expenses) for a poisoned family member or pet!

Now wait a minute, some people say, doesn't the water pressure in the system keep the irrigation water from going backwards? Yes, most of the time it does. But there are times when the water pressure drops in the supply system, and this is when the backflow occurs. No, it is not a frequent occurrence. But it does happen more often than you think. Such as when they have to shut off the water to repair a water pipe, or hook up a new pipe. This makes construction projects easily the most common cause of backflow problems. Fire fighting is another common cause of backflow. Fire trucks use huge pumps to suck the water out of the fire hydrants. This often causes the water pressure in the water system to drop, and backflow will occur in the surrounding neighbourhoods. You can do a quick experiment yourself and create backflow.

Simply tun off the water valve leading to your house. Next have someone turn on a faucet. Now turn on a different faucet that is higher than the first. You will hear air being sucked into the higher faucet. You just created backflow in your house piping. Pretty easy, wasn't it?

There are several types of backflow preventers available, so let's take a look at your options. Always check with the local water provider or building department to be sure that the backflow preventer you plan to use is legal to use in your area. Regulations vary depending on location.

Following is a list of the various types of backflow preventers. All of the following backflow preventer types are available in all sizes. You may need to contact a specialty irrigation store or plumbing supplier to obtain some of these backflow preventers.

Control Valve: A "control valve" is the valve that is used to turn on and off a group of sprinklers or a drip system. Typically an irrigation system will have several control valves, each turning on the irrigation in a different area of the yard. A control valve may be automatic (turned on and off by a timer) or it may be manual.

Atmospheric Vacuum breaker. The atmospheric vacuum breaker (AVB) is the least expensive backflow preventer. The AVB is installed on the pipe right after the control valve. If you use AVBs for backflow prevention you must install one AVB after EVERY control valve, no exceptions. It must be installed at least 6 inches (150mm) higher than the highest sprinkler head, bubbler, or drip emitter outlet that is turned on and off by that control valve (some AVB manufacturers, and in some areas local officials, require that AVBs be installed 12 inches (300mm) higher than the sprinkler heads). As a general rule AVBs are not economically practical if you have more than 6 or so valves. In this case you would want to consider a pressure vacuum breaker. Some municipalities do not allow the use of AVBs. Most people use a anti-siphon valve rather than a valve and a separate AVB. An anti-siphon valve is generally less expensive and less work to install. If you install any valves, of any type, on the pipes downstream of the AVB, the AVB will not work! The downstream valve creates "back pressure" on the AVB which causes the vent in it to jam in the closed position. If this vent can't open, the AVB will not prevent backflow.

Anti-Siphon Valve. A anti-siphon valve is a manual or automatic control valve with a built-in atmospheric vacuum breaker. Like the AVB it must be installed 6 inches (150mm) higher than the highest sprinkler head or drip emitter outlet. If you install any valves, of any type, on the pipes downstream of the anti-siphon valve, the anti-siphon valve will not work! The downstream valve creates "back pressure" on the anti-siphon valve which causes the vent in it to jam in the closed position. If this vent can't open, the anti-siphon valve will not prevent backflow.

Anti-siphon valves are the most common type of backflow preventer used on residential irrigation systems, primarily because they are simple and inexpensive. Some municipalities do not allow the use of anti-siphon valves, so it is best to check with the water company first. Generally you would install the anti-siphon valves in one or more groups, at the highest point in the area to be irrigated. A mainline pipe is run to the anti-siphon valve location(s) from the water source. Pipes then extend from each anti-siphon valve to the sprinklers or emitter tubes. Because anti-siphon valves must be installed at least 6 inches (150mm) above ground, it is a good idea to put a small planting of shrubs around them to help hide them from view. They are not particularly attractive!

Water may come out of the anti-siphon valve periodically, so make sure you install them someplace where a little spilled water will not be a problem. The water will come out of the vent, which is under a cover on the top of the downstream side of the valve (you can see the vent holes under the cover if you turn the valve upside down and look for them.) If water does come out of the anti-siphon it means something is wrong that needs to be fixed. In most cases it means either a stick or rock got into the anti-siphon seal and jammed it open, or the anti-siphon valve was not installed higher than all the sprinkler heads or emitters.

1. Never install a anti-siphon valve upstream of any other valve. If you do the anti-siphon valve will not prevent backflow and you have wasted your money buying it.
2. Never use an anti-siphon valve as a backflow preventer installed on the mainline upstream of other valves. This is a common error that a lot of people make.
3. If you plan to use anti-siphon valves, every one of your sprinkler or drip zone control valves must also be an anti-siphon valve. You can use a ball valve upstream of the anti-siphon valves for an emergency shutoff. But not downstream of them.
4. Anti-siphon valves should never be installed below ground.
5. You can build an enclosure around the anti-siphon valves to hide them. But they must be above ground, and the enclosure must allow water to freely drain out of it if the anti-siphon valves leak.

Pressure Vacuum Breaker. A pressure vacuum breaker (PVB) is similar to a atmospheric vacuum breaker except that you only need to install one of them and it is installed on the mainline leading to the control valves. Like the AVB it must also be installed above ground and it must be 6 inches (150mm) higher than the highest sprinkler head or drip emitter controlled by any of the valves. In a sloped yard it would typically need to be installed at the highest point in the yard, with a mainline pipe running up to it from the water source, and then another mainline running back down to the control valves.

A few local authorities require that the PVB be installed within 18 inches (450mm)of the connection to the water source, in which case you can't use a PVB unless the water source is at the high end of the irrigated area. Some municipalities do not allow the use of PVBs with drip irrigation systems. Some don't allow the use of a PVB at all, so check with your water provider. A PVB backflow preventer may spit or spill water out from under the cap when backflow occurs, so it should not be installed in a location where water spillage would not cause problems.

Warning: If used on a water system where a pump and pressure tank supplies the water (like is used on most rural homes that have a well), the PVB *may* spit water shortly after each time the pump runs. This is because the pressure variations caused by the pump and pressure tank system can cause backflow from the irrigation system back into the water system. The likelihood of water spitting, and the amount of water that spits out, both increase with a longer mainline on the irrigation system. So if you have 10 feet (3m)of mainline between the PVB and the farthest valve there is less likely to be water spitting than if you have 500 feet (150m) of mainline pipe. One way to stop, or at least reduce, this water spillage is to install a spring-loaded check valve right after the PVB. If the check valve is above ground it should be brass, if underground it may be PVC. The check valve may be buried, but be sure to mark where it is with a stake or something in case you need to dig it up and repair it someday! The PVB may still spill a little water with the check valve installed, however in most cases it should be much less water.

Reduced Pressure Backflow Preventer. The reduced pressure backflow preventer (R.P. Unit) is the king of the backflow preventers, made for high-hazard uses. It is also an expensive piece of equipment. It is the standard for commercial irrigation installations. This is the type of backflow preventer. The R.P. Unit must be installed 12 inches (300mm) above ground, but it does not have to be higher than any of the sprinklers. A single R.P. Unit is installed upstream of all the valves. R.P. Units are used for many things other than irrigation systems. Drive through any commercial business area and you will likely spot a lot of these units, most often sitting right out by the street. Many times you may see several grouped together, each used for a different purpose.

R.P. Units may spit out water if they detect backflow, they also spit water if they are broken. So don't install them inside a building without providing a floor drain.

Double Check Backflow Preventers. Depending on who you ask, double check backflow preventers may or may not be appropriate for irrigation systems. In many communities they are legal to use, and even recommended by local officials. Other communities do not allow them to be used on irrigation systems.

Mainlines

The mainline consists of all the pipes from your water source to your irrigation valves. Your water source may be a water company pipe in the street or alley, a well, lake, stream, tank, or whatever. You will likely have two mainline sections, the existing house mainline which goes from your water source to the point where the irrigation system connects, and the irrigation mainline from the irrigation system connection point to the irrigation control valves. In most areas the requirement for the type of pipe are written into local laws. 2" size and larger must often be CL 315 PVC, but check with local building officials.

If your house mainline is polyethylene (a flexible black plastic which usually has a slightly oily feel to it), then you will probably want to use heavy wall (rated 125 PSI) polyethylene pipe for the mainline also, however it is not as durable as PVC and you should check local codes to make sure it is still permissible. Normally the pressure rating of the pipe should be double the actual water pressure. You can also use copper or galvanised steel pipe for new mainlines if you wish. Copper is expensive, but a great choice as it is very durable. Galvanised steel is not such a great choice, it corrodes from the inside and will slowly choke off the flow over time. Galvanised steel also flakes off bits of rust which can cause problems with irrigation systems.

In some cases you may have two (or more) different pipe sizes or types between the water source and irrigation valves. For example, you may have a 1" copper house mainline and a 1 1/4" PVC irrigation mainline. Often it is advantageous for the irrigation mainline to be larger than the house mainline. In any case, you will need to calculate the pressure loss separately for each different pipe size and type, then add them all together.

Example: You have a 1" polyethylene "house mainline" from the water source (a large water company pipe in the street that supplies water to all the houses in the neighbourhood). This polyethylene pipe runs to the water meter and then continues on to the house. A 1" copper mainline then runs through the basement wall to a backflow preventer in the basement and then the copper goes back outside again (this copper pipe is also part of the "house mainline"). Finally you are planning to add a new 1 1/4" polyethylene "irrigation mainline" running from the copper pipe to your irrigation valves. You will need to calculate the friction loss for each of those sections of pipe separately, then add them all together for the total mainline pressure loss.

1" poly house mainline to meter and then continuing to house + 1" copper mainline + 1 1/4" poly irrigation mainline = Total mainline pressure loss

If the water company pipe is somewhere out it the street, how do you know where it is so you can measure the distance from it? You will probably need to make an educated guess. Look up and down the street for manhole covers labelled "water". Generally they are located directly over the pipe, and the pipes in most cases run parallel to the street curbs. Globe valves can be installed almost anywhere you want, although in most cases it is best if they are near (but not within) the area that they will control the water of. Anti-siphon valves need to be at the highest area of the yard. Perhaps an example will help!

So let's say you have a typical house where the water supply pipe comes from the street and runs to the house. You decide to tap into the water supply for the irrigation system in the basement, right after the pipe comes in from the yard.

So your point of connection is in the basement. From there the irrigation mainline would run out through the basement wall to the front yard. The part of the irrigation mainline that is within the basement should be metal pipe. Many people install their backflow preventer in the basement also. That is what we will be doing for our example. So we will put a reduced pressure backflow preventer on the new mainline before it leaves the basement. From the backflow preventer the metal irrigation mainline pipe continues through the wall and out to the yard. Be sure to seal the hole through the wall around the pipe! After the irrigation mainline is through the wall the metal pipe should extend another 12 inches beyond the edge of the wall, then you can change to plastic pipe. If the pipe comes out of the wall above ground it will need to be metal until it is underground. Plastic pipe should never be used above ground.

You plan to water the front yard with the new sprinkler system, so you will need some control valves for the front yard. If you are using anti-siphon valves you don't have a lot of choice as to where they are installed. They must be installed at the highest point in the front yard. In this case we are using a backflow preventer and globe valves, so we don't need the valves at the highest point. Where the new mainline comes through the wall there is already a big group of shrubs, perfect to hide the valves in. So the mainline will come out of the basement and extend to the front yard valves, which are located in a clump of shrubs. So far, so good. Now you also want to water the backyard. You could put the valves for the backyard in the same clump of shrubs with the front yard valves. But this would mean lots of pipes going to the backyard from the front.

Plus it would be inconvenient- if you needed to work on the backyard sprinklers you would have to walk all the way around to the front yard each time you wanted to turn on the valve. So the best solution is usually to continue the mainline from the front yard, around the house, to the back yard. Now the question is where to put the valves for the backyard? If you plan to use anti siphon valves you must install them at the highest point in the back yard. So your choice is already made for you. If you plan to use a backflow preventer and globe valves, then the choice of valve locations is up to you. Again, a good location is someplace out of the way, but also near the area they will control the water in. A good place for the valves is along the perimeter of the yard. Usually a shrub border is planted at the perimeter of a property, so this also works good to hide the valves.

If you need to manually turn on the valve you are going to get sprayed by the sprinklers when you turn them on. This is not fun. Then you need to calculate the total pressure loss in your mainlines and add it to all of your other pressure losses in the sprinkler system. If the pressure losses are too great, you may then need to increase the size of your new mainline. More on that later. Unfortunately, there are no "magic" ways to determine what size a pipe should be. Pipe size is based on many different variables, what works at one house may not work next door!

Almost always the irrigation mainline will need to be the same size or larger as the house mainline, so that is one good place to start. Even better, start with one size larger than the house mainline as in most cases this will be needed, especially if your Design Pressure is less than 50 PSI. Another good starting point is one size larger than the water meter if you have a meter.

Remember that with sprinkler system pipes bigger pipe almost never hurts anything and is almost always better. Using a larger size pipe will not hurt anything except in a few very, very rare situations. The chance of you actually having one of those situations is close to zero. On the other hand, using a smaller pipe can hurt in many situations. When you make the pipe smaller, and force the same amount of water through it, the water must travel faster to squeeze through that smaller pipe. Bernoulli's Principle states that as the speed of a moving fluid increases, the pressure within the fluid decreases. Decreasing the size of the pipe will not create more water pressure.

You can test it yourself- go to Mark Mitchell's Animated Demonstration of Bernoulli's Principle where there is a fantastic, free, interactive graphical demo that allows you to change the size of a pipe and see what happens to the velocity and pressure.

It is better if you avoid running the irrigation water through the house. Connect to the house mainline between the water source and the house if possible. If you do need to install a backflow preventer in the basement to keep it from freezing, tap into the water supply as close as possible to the point where it enters the basement. You can then run a new pipe for the irrigation under or around the house if you want to get the water to the back yard. Most existing house pipes are too small to handle the volume of water needed for an irrigation system. Plus the noise of the water flowing through the pipes may drive you nuts!

VALVES

Type of Valve

There are many different kinds of valves available. You will need at least two different types for your irrigation system.

An emergency shut-off valve. This valve should be installed at the closest point possible to your water source, that is, the location where you tap in for the irrigation system. Without this valve you will need to shut-off the water to the entire house when you want to work on the mainline or irrigation valves. The most commonly used valves for this purpose are "gate valves" because they are inexpensive. Unfortunately the cheap gate valves you're likely to use also tend to not close completely and fail after a short period of time. These cost a bit more but are much more reliable and will last several times longer. So if you pay twice as much for a ball valve it's probably still the best deal! If you do use a gate valve make sure that it is a "wedge" type and buy a good quality one (it will cost as much or more than the ball valve, but it will also work as good). There's nothing worse than trying to work on a system when you can't shut off the water completely.

Control valves. These are the valves that turn on and off the sprinklers. You have two basic styles to choose from.

Globe or angle valve. This valve is available in any size and is commonly installed underground in a box or vault. Since it doesn't incorporate a backflow preventer you must provide one separately. The globe style valve is the most commonly used valve on commercial sprinkler systems.

Anti-siphon valve. Available only in 3/4" and 1" size. The anti-siphon valve incorporates a backflow preventer into the valve. This saves a considerable amount of money, as backflow preventers are very expensive. The anti-siphon valve must be installed above ground and must be at least 6" higher than the highest sprinkler head. This means that if you want to use anti-siphon valves you will have to locate the valves at the highest point in your yard, and run a water supply pipe to them from the water source. The mainline should be buried 18" deep.

Operation Method

The control valves may be manually operated or they can be remotely controlled. Manual control is simple, the valve has a handle you use to turn it on. Remote control valves are either electric or hydraulic, but almost everyone uses electric. The electric valve operates on 24 volt alternating current (vac) and is turned on and off by a timer called an "irrigation controller" or often just "controller". Anti-siphon, globe, and angle valves are all available as automatic valves. Pretty much all valves and controllers are compatible unless the controller is battery or solar operated. So you can buy a brand "X" controller and it will work fine with brand "Y" valves. Or you can even use various brands of valves together.

Flow Control

If you are going to use automatic valves, you select one that has a manual flow control on it. This is in addition to the manual on and off control on the valve. The flow control bypasses the automatic valve features allowing the valve to be closed by turning a handle just like a standard manual valve. It also allows the valve to be "throttled", that is, left in a partially closed status.

— You can manually force the valve closed if it sticks open. Failure to close automatically is the most common valve problem, so there's a good chance that someday you will use the flow control.

— If your flows are low, it may be necessary to throttle down the flow control to make the valve close reliably.

— Partially closing the flow control will make the valve close faster, which is not something you want to do normally, but sometimes it is necessary. On automatic systems it is common for the next valve to open before the previous one fully closes. The resulting loss of pressure due to two valve circuits being on at the same time can cause the first valve to never close. This is often a problem with systems that use pumps to supply the water.

— You can use the flow control to throttle the flow if your sprinklers are misting or your pressure increases for an unknown reason.

— If a sprinkler breaks and you don't have time to fix it, you can shut off the valve with the flow control and all the other valves will still operate as normal. That way you don't have to mess with the controller programme.

Materials

Valves come in both brass and plastic models. Most used today are plastic, but brass is not at all out of the picture. There is no doubt that a brass valve will last longer if installed in the sunlight. From an operational point of view both are reliable, especially for automatic systems. For manual valves my experience is that brass will last much longer. If you use plastic valves above ground you may wish to consider building a cover for them to protect them from sunlight, which can destroy the plastic over time. Two types of plastic material are used for valves. Glass-reinforced nylon is the best, it is tougher, more resistant to impact, and has a higher pressure rating. PVC is used for lower cost valves, it still is as strong as the pipe you're using, although that really depends on how thick the plastic is! If the valve fails, they can be difficult to replace. Plus only the cheapest valves come with solvent weld connections, so what does that tell you about the likelihood you will need to remove it some day?

Maintenance

Today's valves are pretty maintenance free. Almost all automatic valve failures result from installation or design problems. Ignore the following and you will hate your valves regardless of what type or brand you buy.

- Dirt in the irrigation pipes. Inside the valve there are very small water passages that lead to and from the solenoid. If a grain of sand or glob of algae gets into these passages it can block them and the valve will fail to open or (more likely) fail to close. It is critical to flush all the dirt out of the pipes before installing the valves. A 150 mesh filter installed at the water source connection can also help keep out sand that comes in with the water supply. You may be surprised to learn that most water companies have considerable amounts of sand in their pipes. When you install a new sprinkler system the higher flows stir up this sand and then it gets into your new system.
- Almost all valve solenoid failures are caused by water getting into the solenoid. The water gets into them through the wires. The solenoid wires have multiple strands of wires twisted together with insulation around them. Because they are twisted there are very small gaps between the wires. Water is sucked up through these small passages into the solenoid by capillary action. Thus it is critically important that the

wire splices on the valves be completely water proof so that water can't be sucked into the solenoid through the wires. You should water-proof the wire splices right after you test the valves! They only need to get wet once to ruin the solenoid. The Installation Tutorial has more on this.

Valve Size and Pressure Losses

Emergency shut-off valve

The pressure loss through the emergency shut-off valve is not significant enough to worry with. We will ignore it. The emergency shut-off valve should be the same size as the pipe it is installed on.

Automatic valves

The pressure loss in an automatic valve is the energy source used by the valve to open and close. If the valve doesn't have enough pressure loss it will not have the energy needed to close by itself. Always size automatic valves based on the flow rate using the manufacturer's chart as a guide. Never assume that the valve should be the same size as the pipe! It is very common for the valve to be a different size than the pipe it is installed on. If you absolutely must guess, use the next valve size smaller than the pipe size and assume a pressure loss of 6 PSI. Never guess if your flow is less than 5 GPM, always use a chart! Many automatic valves won't work at all at flows below 5 GPM!

The size of the automatic valves is determined by the manufacturer's recommended flow range, together with the pressure loss through the valve at the selected flow. You will need to get the valve manufacturer's flow chart for the model of valve you plan to use. This information should be on the valve package.

For manual valves

Manual valves are much more forgiving than automatic valves. You don't need to worry about having enough pressure to allow the valve to close by itself, it uses "elbow grease" to power it! However, you still need to find out what the pressure loss through the valve will be so you can enter it in your Pressure Loss Table. As with the automatic valves, you use a chart provided by the manufacturer for this. Follow the same procedure given above for automatic valves. Unfortunately, pressure loss data for manual valves can be hard to find as many manufacturers don't provide it. As a general rule, allow 2 PSI pressure loss for a globe or angle type manual control valve, 5 PSI if it is an anti-siphon valve.

Manual irrigation control valves should be of the "angle" or "globe" type with replaceable rubber seats. Never use a gate valve as a control valve. It is not made to be regularly opened and shut. Many gate valves will fail after as little as 10 uses! .

Sprinkler Heads

Types of Sprinkler Heads

Sprinklers heads are divided into two types based on the method they use to distribute the water.

Spray Heads: More properly called "fixed spray heads" these are the small heads that spray a fan-shaped pattern of water. Think of a shower nozzle. Most use interchangeable nozzles installed on the sprinkler which determine the pattern (1/2 circle, full circle, etc.) and the radius of the water throw. Some specialty patterns are available for long, narrow areas. Spray heads are spaced up to 18 feet apart. They need between 20 and 30 PSI of water pressure to operate properly.

Rotors: Rotor is the term used to describe the various sprinklers which operate by rotating streams of water back and forth over the landscape. The example which most people are familiar with is the "impact" rotor sprinkler which moves back and forth firing bursts of water. You probably know this sprinkler best for the distinct sound it makes when operating— tooka, tooka, tooka, tic, tic, tic, tic, tic, tooka, tooka, tooka, etc... The impact rotors are rapidly being replaced now by gear driven rotors which are very quiet, lower maintenance, and much smaller in size. These gear drive rotors have one or more fingers of water which move silently across the landscape. The prettiest of these are the "multi-stream rotors" where multiple streams of water rotate over the landscape one after the other.

Multi-stream rotors are fascinating to watch but tend to be more expensive and much, much, much higher maintenance than the other gear drive heads. Rotors can be spaced from 18 feet to 55 feet apart. Rotors require a lot more water pressure to operate than spray heads. The water pressure at the rotor head in (PSI) must exceed the distance (feet) between the heads. Thus if you want to space rotors 35 feet apart you will need 35 PSI of pressure at the rotor. More on this later. The small rotors sold for residential use work best at 25 to 35 foot spacings.

Which to use, sprays or rotors? Rotors are spaced farther apart and require less trenching, but cost much more per sprinkler. If the area to be watered is less than 18 feet wide then you should use spray heads. For most normal-size city residential yards spray heads are the better choice. For areas less than 22 feet wide spray heads are generally less expensive. In areas larger than that it comes out about the same cost for rotors or spray heads.

Basic Body Styles

Pop-Up Style Sprinklers: Pop-up style sprinklers are installed below ground. A portion of the sprinkler rises up out of the ground when the sprinkler is operating and then retracts back below ground when not in use.

Shrub Style Sprinklers: Shrub style sprinklers are installed above ground on top of a pipe called a "riser".

In most cases you will want to use pop-up style heads, even in shrub areas. Pop-up sprinklers often don't cost any more than shrub sprinklers when you include the cost of the riser (the upright pipe the sprinkler is mounted on). Two major advantages of pop-up sprinklers are safety and appearance. Many people are injured each year when they trip over, or fall on, shrub style sprinklers. For shrub and ground cover areas pop-up sprinklers are available with pop-up heights of 3", 4", 6" and 12" above ground. A few brands are available with lower pop-up heights, but be warned that lower heights often cause problems. As a general rule it is best to also avoid shrub style sprinklers unless a very tall riser is needed to raise the sprinkler spray over the tops of tall shrubs. When needed, shrub style sprinklers should only be used in areas well away from sidewalks, patios, and areas where children play.

Metal or plastic

At the grocery store it's "paper or plastic?" but with sprinklers the question becomes "metal or plastic?". The conventional wisdom is that metal is more durable than plastic, and therefore is better. Up until the late 1970's metal (usually brass, sometimes zinc) was the standard material from which almost all sprinklers were made. However, times have changed and now plastic is the most common material for sprinklers. Very few manufacturers even bother to make an all-metal sprinkler anymore. The primary reason for this change in materials is cost; machined metal parts are enormously expensive in comparison to injection molded plastic. Fortunately, most of today's plastic sprinkler heads are very well engineered and will perform as well as, if not better than, the old metal sprinklers.

Hybrids: A few companies manufacture plastic sprinkler bodies which accept brass nozzles, which they claim results in a better water pattern. Other manufacturers claim that plastic nozzles perform as well as brass. The research tends to indicate that a really well-machined brass nozzle has better water distribution. But that's laboratory tests, and in the real world a lot of other factors come into play. More importantly, there are a few nozzles, both brass and plastic, which don't seem to perform as well as others. Fortunately, they are easily identified by comparing prices.

Features

The following features are common to all good-quality sprinkler heads (for both rotors and spray type heads). Choosing a sprinkler without these features is asking for trouble.

Spring retraction: Make sure a spring is used to pull the pop-up riser (sometimes called a "piston") down into the case when the sprinkler isn't on. As a general rule the stronger the spring, the less likely the riser is to "stick up". Stay away from sprinklers that rely only on gravity to retract the pop-up riser.

Wiper seal: This is a soft plastic seal around the pop-up riser stem that seals the riser so it won't leak . The wiper seal also is responsible for keeping dirt out of the sprinkler body, and is the most important part in determining how long the sprinkler will last. Make sure the sprinkler model you select has a wiper seal. Note: on some sprinklers you must remove the sprinkler's cap and look inside the bottom of it to see the seal. Be careful when removing the cap, on some models the spring will shoot out!

3 Inch Pop-Up Height (or higher): Unless you just like to trim grass around sprinkler heads, make sure the pop-up height is 3" or more. This way the spray nozzle will clear the top of the grass. Most professionals use 4" pop-up sprinklers in lawn areas, and 6" or 12" pop-ups in shrub areas.

Pressure Requirements for Sprinklers

Spray type sprinklers

For spray type sprinklers most designers use an operating pressure of 30 PSI, unless a lack of available pressure forces a lower level. Remember that if you use a lower pressure the sprinklers will need to be spaced closer together, because the water won't spray as far. Sprinkler manufacturers provide charts that tell you how much pressure is required for the sprinkler and how far it will spray with that pressure. Additionally, almost all spray type heads have a radius adjustment screw that allows you to adjust the watering radius down for smaller areas. At pressures above 45 PSI most spray heads start to "mist" which results in poor irrigation. This can be controlled by using the radius adjustment feature to reduce the pressure, installing a pressure regulator on the mainline, or by using special pressure regulating nozzles made by some sprinkler manufacturers.

Rotor Type Sprinklers

For rotor type sprinklers the higher the operating pressure the better. Most rotor type sprinklers do not work well with less than 30 PSI operating pressure. Stryker's Rule" states that the spacing in feet between rotor-type sprinklers can't exceed the pressure in PSI at the rotor. Do not believe the manufacturer's literature on rotor radius! They get those distances by testing the rotors inside a big building with no wind. Even the most gentle breeze will shorten the radius (water droplets are very light). If you ignore this rule you will have dry spots! So if you want to space rotors 30 feet apart you will need to allow for 30 PSI of pressure at the sprinkler head. If you want to space rotors 40' apart you will need 40 PSI for the sprinkler head pressure.

Drip Emitters

Most emitters operate best at around 20 PSI. Some emitters are "pressure compensating" which means they should put out approximately the same amount of water over a wide range of pressures. Keep in mind that at pressures over 45 PSI emitters may blow apart. Barbed emitters in poly tubing may pop out of the tubing at 30 PSI.

Mix and Match

Sometimes you need to use sprinklers that require high pressure such as rotors, with sprinklers that use low pressure on the same irrigation system. To do this the system is designed using the pressure requirements of the high pressure sprinklers. The low pressure sprinklers (or emitters) are installed so that a separate valve turns them on and off, and a special pressure reducing valve is used. This valve reduces the pressure down to the correct amount for the low pressure sprinklers. Almost all irrigation manufacturer's now make pressure reducing valves, although you may have to go to a specialty irrigation store to get them.

Remember- the pressure you enter in your table is the pressure for a single sprinkler head. So if you will have 10 sprinklers and they each require 30 PSI you still only write "30 PSI" on your pressure loss table. Also the value you enter should be the highest sprinkler head pressure requirement. So if you plan to use a spray head that will need 20 PSI and also a rotor that will need 35 PSI, you will enter the higher value- which in this case would be 35 PSI. Finally, remember why pencils have erasers. You can always come back and change this value later if you want to! So don't agonize over it.

Lateral Pressure Loss

Types of Commonly used Irrigation Pipe

Two types of pipe are commonly used for sprinkler systems laterals, polyvinyl chloride (PVC) and Polyethylene (Poly). Both are types of plastic. PVC is usually white or gray colour and semi-rigid. Polyethylene is usually black and is flexible.

— PVC is the type most commonly used in warm winter climates. PVC pipe is rated by two different systems, the first is the "class" system (Cl) the other is the "schedule" system (SCH). It is not possible to say that one is always better than the other. Schedule pipe is rated by the pipe's wall thickness, while class pipe is rated by the pipe's operating pressure. All PVC pipe of the same size will have the same outside diameter, regardless of which type or rating they are, so the same fittings will fit all of them. Most PVC pipe is connected together using "PVC fittings" which are glued in place. The fittings are typically rated as SCH 40 (standard white PVC fittings), some are available as SCH 80 (stronger and normally gray colour). Sometimes PVC pipe has threaded ends just like steel pipe. PVC pipe and steel pipe have the same outside diameters, and threaded steel pipe fittings will fit on threaded PVC pipe.

— Poly pipe is commonly used in areas with very cold winters, and where special situations occur such as very rocky soil. Poly pipe doesn't break as easily if water freezes in it, but also doesn't hold up as well under higher pressures. Poly pipe is more forgiving in rocky soils (big rocks, like granite boulders) as it is less likely to crack if it is installed against the side of a large rock. Poly is rated by a system of "SDR" ratios. The lower the SDR number, the stronger the pipe is. All poly pipe of the same size will have the same INSIDE diameter. This is because poly pipe uses "insert fittings" also sometimes called "barbed fittings". Insert fittings shove into the pipe and have barbs to hold them in place. Do not rely on the barbs! Use a stainless steel clamp to hold the fitting in place.

Hard Frost Areas

In areas with hard frost pipes should be installed below the frost line or you should have drain valves installed at all low points to allow them to be drained in the winter. You can also have the pipes winterized by blowing the water out of them with air for the winter season.

Laterals

"Lateral pipe" is the name given to the pipe sections between the control valves and the sprinkler heads or drip emitters. If you visit your local warehouse hardware store you may find that they use the name "branch" pipe for laterals. We will determine the actual sizes of these pipes later using the pressure loss value that we establish here. For most residential sprinkler systems on City size lots a lateral pressure loss value of 4 PSI will work great. It gives a nice balance between cost savings and performance and is a safe figure to use with any type of sprinkler. If you decide you want to use a lower value you can do so without problems.

Sprinkler Risers

Sprinkler risers are the short pipes that connect the sprinkler heads to the laterals. They are usually flexible to allow the sprinkler head to move without breaking the lateral pipe. That's a nice feature to have when you drive over a sprinkler with the car or hit a sprinkler with the mower. A standard riser would be a PVC swing joint using pipe and fittings the same size as the sprinkler inlet, any of the poly pipe risers commonly know as "Funny Pipe", "swing pipe" or equal (keep the riser pipe less than 18" long), or Cobra Connectors (keep them less than 12" long).

Sprinkler Spacing

Sprinkler Coverage

The area watered by each sprinkler must overlap substantially the area watered by the adjacent sprinkler. This overlap may seem like a waste at first, but it is a very important necessity. Without this overlap it would be impossible to design sprinkler systems that provided uniform water coverage.

Sprinklers are intentionally designed to require 100% overlap of watered areas. That means each sprinkler throws water ALL the way to the next sprinkler in each direction. That's right, 100% overlap of watered areas is required or you will get dry spots! This is known in the industry as "head-to-head coverage or head-to-head spacing". A lot of those free design brochures you find in stores get this wrong. They don't show enough overlap! The writers of those brochures think you are going to look at the overlap and buy the brand of sprinkler that shows the least heads. So they try to make it look like you can use less heads with their sprinklers. After you've bought the sprinklers if you have dry spots, well hey, it's your problem now! You'll probably just buy a few more sprinklers to fill in the gaps.

Sprinkler Precipitation Rate and GPM

The precipitation rate is the amount of water the sprinkler throws onto the area it waters, measured in inches per hour. (Inches per hour is how deep, in inches, the water would be after one hour if it didn't soak into the ground or run-off.) Precipitation rate must be considered when selecting your sprinkler heads to eliminate water application uniformity problems (dry spots).

Spray heads: Almost all sprinkler manufacturers make their spray heads so that you can mix and match nozzle patterns and the precipitation rates will still match for all the heads. This is referred to as "matched precipitation rates". Look for this feature when selecting your sprinklers. Important: do not mix different brands of spray heads and nozzles together on the same valve circuit without checking to see that they have the same performance specifications. Just because the nozzle will screw into the sprinkler body doesn't mean it's designed to work with that sprinkler!

Rotors: Rotor-type heads aren't quite as easy. You must select the appropriate nozzle size for each rotor in order to match the precipitation rates. A simple illustration will help explain. Rotor heads move back and forth across the area to be watered. The rotation speed is the same regardless of whether the rotor is adjusted to water a 1/4 circle or a full circle. So the stream from a 1/4 circle head will pass over the same area 4 times in the same amount of time that it takes for a full circle head to make one pass over the area it waters.

With the same size nozzle in both, a 1/4 circle rotor will put down 4 times as much water on the area under the pattern as a full circle rotor will. To match the precipitation rates between these sprinklers, the quarter circle rotor must have a nozzle that puts out 1/4 the amount of water that the full circle nozzle puts out! A half circle rotor must have a nozzle that puts out 1/2 the water of a full circle. This is why when you buy a rotor-type sprinkler head they often include a handful of different size nozzles with it.

If you have rotors that are adjusted for different radii you will need to adjust the nozzle size to compensate for the radius change also! For example if most of the rotors are set for a 30 foot radius, but one is adjusted down to 20 ft., the 20 ft. one will need a nozzle 1/2 the size.

References

Norum, E.M., "Sprinkler System Returns Wastes to the Land", *Published in World Farming magazine,* June, 1973.

______________ ., "Special Design Considerations for Center Pivots", *Paper presented at Sprinkler Irrigation Association Annual Technical Meeting,* February, Kansas City, Kansas, 1976.

______________ ., "Turf Sprinkler Performance Data", *CIT Technical Note Series,* 1984.

4

Design of Travelling Irrigators

Travelling irrigators or continuous-move irrigators are systems, which apply water while they are in motion. They include the cable drawn, also known as hose-drag irrigators, and hose reel, also known as hose-pull rain gun irrigators as well as the centre pivot systems. All systems were developed in the early 1970s, because of an increasing cost of labour. The hose-drag irrigator was developed in the USA while the hose-pull irrigator was developed in Europe, mainly for supplementary irrigation.

A travelling irrigator consists of the following major components: pumping unit, mainline, flexible hose, traveller unit and a gun sprinkler or a series of impact sprinklers or spray nozzles mounted on folding booms. This module describes the two types of hose-fed travel irrigators, namely the hose-drag and the hose-pull irrigator. Their principle of irrigation is the same. They differ in the mechanism with which the hose operates.

The traveller for hose-drag systems consists of an irrigator chassis on which a winch is mounted. The irrigator chassis also carries the irrigation equipment, such as the rain gun or boom sprinklers. The winch is driven by a piston and a ratchet mechanism, which is powered by water, bled off from the main supply to the irrigator itself. Water is supplied to the cable drawn irrigators from a hydrant on the mains via a lay flat hose. The lay-flat hose, laid in a loop at the start of irrigation, is dragged as the machine moves. Therefore, a loop carrier is normally provided in order to reduce crop damage. The cable has to be run out to the other end of the field where it is anchored. During irrigation, the irrigator winds itself across the field while dragging the hose towards the place where the cable is anchored. The cable should be equal to the tow-path in length.

The traveller for the hose-pull consists of a rain gun mounted on a sledge. The gun is pulled by means of its own hose which is connected to a large drum (hose reel) on the other end of the field. The hose is pulled as it winds on the drum. The drum is mounted on a chassis, which is parked at the headland, that is the water supply end. The chassis is fitted with a drive mechanism, which in the majority of cases is driven by water-power from the main supply (turbine or piston or ratchet). Some of these travellers may have the reel on the sprinkler end of the hose. The hose reel should be strong and capable of accommodating the full length of the hose while it is full of water and under high pressure.

More recently, the folding boom equipped with spray nozzles has been gaining popularity in view of its low energy requirements. These systems, while maintaining the basic features of hose-reel guns, have substituted the gun with the boom and nozzles. They can operate with pressures as low as 25-35 metres at the mainline hydrant.

Preliminary Design Steps

The general sprinkler design principles such as depth of water application, optimum application rates, system capacity and others are also applicable to the design of the travelling sprinkler. These are the system layout, sprinkler and traveller selection, towpath spacing, the relationship between traveller speed and rate of water application and the frictional losses in the hose and traveller.

The net depth of water application and irrigation frequency can be calculated as follows:

$$d_{net} = (FC\text{-}PWP) \times RZD \times P \text{ (mm)}$$

$$d_{gross} = \frac{d_{net} \text{ (mm)}}{E}$$

$$\text{Irrigation Frequency (IF)} = \frac{d_{net} \text{ (days)}}{wu}$$

The preliminary system capacity (Q) can be obtained using the equation:

$$Q = \frac{K \times A \times d}{f \times T} \qquad (1)$$

Where:

Q = system capacity (l/s)
K = a conversion constant, 2.78
A = design area (ha)
d = gross depth of water application (mm)
f = irrigation frequency or operation time allowed for completion of 1 irrigation (days)
T = irrigation time (hours)

Final Adjustment of the Design

As was done for periodic-move systems, once the preliminary design parameters are established, the final adjustment of the design can be done, starting with the sprinkler selection.

Sprinkler Selection

The sprinkler characteristics that have to be considered as selection criteria are:

— jet trajectory
— operating pressure
— sprinkler body design

The sprinkler operating conditions to be considered in sprinkler selection are:

— soil infiltration characteristics
— desired irrigation depth
— desired or appropriate irrigation cycle
— crop characteristics
— wind conditions
— tow-path spacing

Jet Trajectory

Most travellers use gun sprinklers with trajectory angles ranging between 18 and 32 degrees. High angles give maximum coverage only under low wind conditions and this minimises droplet impact. For winds exceeding 16 km/hour, gun sprinklers with trajectory angles between 20 and 21 degrees should be used. Where winds are below 16 km/hr, sprinklers with trajectory angles from 26 to 28 degrees are better. Low angles generally result in large droplets which are not good for some crops, especially leaf crops such as tobacco.

Sprinkler Nozzles

Generally gun sprinkler nozzles are fitted with either tapered or orifice nozzles. Jets produced by tapered nozzles have large drops and are less susceptible to wind than those produced by orifice nozzles. Since orifice nozzles produce smaller drops, they are used on more delicate crops, which are susceptible to large drop impact. However, the jets from orifice nozzles are also more susceptible to wind.

Sprinkler Wetting Patterns

Gun sprinklers can apply water to the full circle or part of the circle. The choice depends on considerations of uniformity of water distribution and the need sometimes to leave a dry area along which the traveller moves. In general, part circle coverage has a higher uniformity than full circle water application. This is due to the fact that for the same sprinkler discharge, the part circle sprinkler has a higher application rate at each point. As an example, half circle coverage doubles the full circle application rate, for the same sprinkler under similar conditions. Therefore, half the time would be required to apply the same depth as that applied by full circle coverage.

With these considerations in mind, a sprinkler can be chosen from Table 1 for the design discharge of 24 l/s. Similar tables can be obtained from manufacturers. From Table 1, a gun sprinkler with a nozzle diameter of 30.5 mm and operating at 56.24 m head will discharge 24 l/s. This in fact is the required flow. The diameter of coverage of the sprinkler at that pressure is 120.4 m.

It is important to check whether the precipitation rate of the sprinkler, selected based on the required flow, is acceptable with regards to infiltration rates. The precipitation rate should not exceed the infiltration rate. Equation 2 is used to approximate the application rate of the sprinkler in order to compare it with the infiltration rate of the soil:

$$I = \frac{K \times Q_{sprinkler} \times 360}{\pi \times (0.9 \times R)^2 \times w} \tag{2}$$

Where:

I = approximate infiltration rate or approximate sprinkler application rate (mm/hr)

K = conversion constant, 3600

Q = sprinkler discharge (l/s)

π = 3.14

R = wetted radius of sprinkler (m)

w = portion of circle receiving water (degrees)

This equation assumes that 90% of the radius receives sufficient water.

Full Circle Operation

For full circle operation for the selected sprinkler:

$$I = \frac{3600 \times 24 \times 360}{3.14 \times (0.9 \times 60.2)^2 \times 360} = 9.37 \text{ mm / hr}$$

Table 1. Discharges and wetted diameters for gun sprinklers with 24 degree angles of trajectory and tapered nozzles operating when there is no wind

Sprinkler	Diameter of Tapered Nozzle (mm)									
Pressure	20.3		25.4		30.5		35.6		40.6	
	Sprinkler Discharge and Wetted Diameter									
m	l/s	m	l/s	m	l/s	m	l/s	m	l/s	m
42.18	9.02	86.87	14.20	99.06	20.82	111.25	-	-	-	-
49.21	9.78	91.44	15.46	103.63	22.40	115.82	30.29	132.59	-	-
56.24	10.41	94.49	16.41	108.20	23.98	120.40	34.50	138.68	42.59	146.30
63.27	11.04	97.54	17.35	111.25	25.56	124.97	34.39	143.26	45.12	150.88
70.30	11.67	100.58	18.30	114.30	26.82	128.02	36.28	146.30	47.64	155.45
77.33	12.30	103.63	19.25	117.35	28.08	131.06	38.18	149.35	48.85	158.50
84.36	12.94	106.68	20.19	120.40	29.34	134.11	39.75	152.40	52.06	163.07

Part Circle Operation

If the sprinkler is operated as part circle, in which case it leaves a dry area for the traveller, say 75 degrees, the wetted part of the circle would be 285 (360-75) degrees.

$$I = \frac{3600 \times 24 \times 360}{3.14 \times (0.9 \times 60.2)^2 \times 285} = 11.8 \text{ mm / hr}$$

Having established the approximate application rate of the sprinkler one should check in Table 2 to see whether this application rate is compatible with the soil and topography. If not a different application rate and pressure combination or sprinkler would have to be selected.

Table 2. Suggested maximum sprinkler application rates for average soil, slope and tilth

Soil Texture and Profile	Slope			
	0-5%	5-8%	8-12%	12-16%
	Maximum Application Rate (mm/hr)			
Coarse sandy soil to 1.8 m depth	50	38	25	13
Coarse sandy soil over more compact soil	38	25	19	10
Light sandy loam to 1.8 depth	25	20	15	10
Light sandy loam over more compact soil	19	13	10	8
Silt loam to 1.8 m	13	10	8	5
Silt loam over more compact soil	8	6	4	2.5
Heavy textured soil or clay loam	4	2.5	2	1.5

The sprinkler application rates in Table 2 should be reduced by 25% for traveller sprinklers in order to avoid the erosion of the soil, which could result from the relatively large water drop sizes. This table shows that the sandy loam soil can tolerate a maximum sprinkler application rate of 25 mm under periodic-systems. In order to avoid erosion under traveller irrigation systems the application rate should be reduced by 25% which gives approximately 19 mm/hr. Therefore, both the part circle and full circle sprinklers could be used. If orifice nozzles are used, they would generally reduce the wetted radius by about 5%. This is because orifice nozzles discharge smaller drops, which result in a smaller radius of throw. The resultant application rates would be higher, calling for checking this application rate against Table 2. In our example the wetted radius is 60.2 m (120.4/2). Therefore the wetted radius for orifice nozzle is 57.2 m (0.95 x 60.2 = 57.2). In that case the approximate application rate for, say a part circle, 35 mm diameter orifice nozzle operating at 56.24 m is, using Equation 2:

$$I = \frac{3600 \times 24 \times 360}{3.14 \times (0.9 \times 57.2)^2 \times 285} = 13.1 \text{ mm / hr}$$

This application rate is still acceptable since it is still below 19 mm/hr.

Tow-path Spacing

Tow-path is the path along which an irrigator travels. As several paths would be required to cover the total area, the spacing between these paths should be established. For traveller sprinkler irrigators, the uniformity of water application is affected by:

- — wind velocity and direction
- — jet trajectory
- — nozzle type and wetted sector angle
- — sprinkler sector profile and overlap achieve
- — the variations in the sprinkler operating pressure and traveller speed

According to Collier and Rochester, coefficient of uniformity (CU) values for travel irrigators only go up to 70-75% for

Table 3. Typical recommended tow-path spacings for traveling gun sprinklers under various wind conditions, trajectory angles between 23 and 25 degrees

	Wind Speed (km/hr)						
Sprinkle	Over 16			8-16		3.2-8	0-3.2
Wetted Diameter (m)	Spacing as a Percentage of Wetted Diameter 50	55	60	65	70	75	75
	Tow-path Spacing (m)						
60.96	30.48	33.53	36.58	39.62	42.67	45.72	48.77
76.20	38.10	41.76	45.72	49.38	53.34	57.00	60.96
91.44	45.72	50.29	54.86	59.44	64.01	68.58	73.15
106.68	53.34	58.52	64.01	69.19	74.68	79.86	85.34
121.92	60.96	67.06	73.15	79.25	85.34	91.44	97.54
137.16	68.58	75.59	82.30	89.00	96.01	103.02	109.73
152.40	76.20	83.82	91.44	99.06	106.68	114.30	121.92
167.64	83.82	90.05	100.58	109.12	117.35	125.58	134.11
188.88	91.44	100.58	109.73	118.87	128.02	-	-

wind speeds up to 16 km/hr and this is only in the central parts of the fields. It is therefore necessary to select a tow-path spacing which would provide the best possible spacing between any two tow-paths. Table 3 gives an example of recommended tow-path spacings for travellers with gun sprinklers whose angles of trajectory are between 23 degrees and 25 degrees. Where wind speeds exceed 16 km/hr sprinklers with lower trajectory angles, typically 20 to 21 degrees should be used and where winds are negligible 26 to 28 degree trajectories should be used.

In our example, the selected 30.5 mm diameter nozzle discharges 24l/s at a pressure of 56.24 m and has a wetted diameter of 120.4 m. From Table 3 the nearest wetted diameter to 120.4 m is 121.92 m. Since the wind speed is 3.5 km/hr, it falls within the range 3.2-8 km/hr. Two options are available within this wind speed range and a choice of spacing between 70 and 75% of the wetted diameter should be made. Since the wind speed of the case under consideration is closer to the low end of the wind speed range, the higher percentage is adopted.

It should be pointed out that the higher the wind speed, the more the overlap required and thus the lower percentage of wetted diameter is used. Hence the nearest estimated tow spacing as a function of the wetted diameter is 91.44 m. In our example, the wetted diameter is 120.4 m. According to the table, the tow spacing should be 75% of the wetted diameter, that is 90.3 m (0.75 x 120.4). This is near enough to the recommended 91.44 m. A good round figure to use for the tow-path spacing is 90 m.

For uniformity of application purposes, it is desirable that the general direction of the tow-path follows the contour lines. The field length is 630 m. Therefore the number of tow-paths is obtained by dividing the length of the field by the width of the tow-path spacing. In our example this is 7 (630/90). Therefore in order to cover 630 m, 7 tow-paths are necessary for the traveller to complete the irrigation of the full 25.2 ha.

Travel Speed

Generally travellers are designed and set up to travel one tow-path length in about 23 hours for a single shift per day or 11 hours for 2 shifts per day. For 2 shifts, 1 hour is provided for between the shifts in order to allow for the change to the next tow-path.

The original assumption was that the whole area would be irrigated in 1 single day which is not the case. In our example, it was assumed during the calculation of the allowable minimum system capacity that the set time would be 23 hours showed that the number of tow-paths necesssary to cover 630 m width of field was 7. Therefore effectively it takes 7 days to irrigate 25.2 ha, that is, the irrigation cycle is 7 days.

The frequency obtained was 7.8 days assuming 50% allowable soil moisture depletion. If an irrigation cycle of 7 days and an irrigation frequency of 8 days is adopted then the traveller has to irrigate continuously during the peak water requirement period for 7 days with one day off. This would allow for repairs and maintenance as well as cultural practices. In that case, the moisture depletion at which irrigation occurs is obtained as follows:

Gross irrigation depth (d_{gross})

= 8 days x 7.86 mm/day = 62.9 mm per irrigation.

The moisture depletion at the time of irrigation is:

$$= \{(8 \times 5.5)/(115 \times 0.75)\} \times 100\%$$

$$= 51\%.$$

The average gross application depth (dgross) is given by equation 3:

$$d_{gross} = \frac{k \times Q}{v \times W} \qquad (3)$$

Where:

d_{gross} = gross depth of water application (mm)

k = conversion constant, 60

Q = sprinkler discharge (l/s)
v = travel speed (m/min)
W = tow-path spacing (m)

Therefore to compute v the equation is re-arranged to:

$$v = \frac{k \times Q}{d_{gross} \times W}$$

Substituting the values gives:

$$v = \frac{60 \times 24}{62.9 \times 90} = 0.254 \text{ m / min}$$

Therefore the time required to traverse 400 m length of tow-path is given by:

$$\frac{400}{0.254 \times 60} = 26.62 \text{ hours}$$

This would require certain changes to the design, so that we do not exceed 23 hours of operation per day at peak demand. Several options are available:

1. The length of path is reduced so that it is covered within 23 hours. This would imply that the path length is reduced to 350 m (0.254 x 60 x 23) and a strip of land of 50 x 630 m^2 or 3.15 ha would not be irrigated
2. The irrigation cycle and frequency are 7 days, resulting in irrigating the whole area within 7 days and taking the risk of non-availability of time for repairs and maintenance during the peak demand
3. Assuming a larger diameter of nozzle is selected (35.6 mm), let us go through the same process to investigate if less than 3.15 ha of land remains without irrigation. Combine this with an irrigation cycle of 6 days and frequency of 7 days

$$d_{net} = 7 \times 5.5 = 38.5 \text{ mm and}$$

$$d_{gross} = 38.5/0.7 = 55 \text{ mm}$$

Depletion = {(7 x 5.5) / (115 x 0.75)} x 100% = 45%

The sprinkler flow rate Q = 34.5 l/s and the wetted diameter is 138.68 m. The approximate application rate:

$$I = \frac{3600 \times 34.5 \times 360}{3.14 \times (0.9 \times 69.34)^2 \times 285} = 12.8 \text{ mm / hr}$$

This is less than 19 mm/hr and thus acceptable. The next step is to establish the tow-path spacing from Table 3, which is 103.02 m. In our example, the wetted diameter is 138.68 m, which would give a tow-path spacing of 104 m (0.75 x 138.68). This is close enough to the recommended 103.02 m. Hence, with 6 tow-paths (one per day) a strip of land of 0.24 ha ({630 - (6 x 104)} x 400) can be excluded from irrigation, if conformity to speed is obtained.

$$\text{Conformity speed, } v = \frac{60 \times 34.5}{55 \times 104} = 0.362 \text{ m/min}$$

Therefore, the time required to traverse 400 m of length of tow-path is 400/(0.362 x 60)= 18.4 hours. Other options are available through the selection of smaller sprinklers or the same operating at lower pressure. The final choice is up to the farmer as to whether he/she would accept to take the risks and limitations of the one or other option.

Standing Positions, Times and Hose Length

Standing time is the time during which a sprinkler operates while it is stationary. When using travelling irrigators, only the central positions of the field receive adequate irrigation. A portion equal to the wetted radius of the sprinkler, on each end of the field, receives inadequate irrigation. This will always be the case, but can be minimised by allowing for standing time at each end of the tow-path, so that those areas can receive more water than would otherwise be the case. It should be kept in mind however, that in this case the opposite to the edge area will receive more than the needed amount of water. The standing positions and time depend on:

— whether the traveller moves from end to end
— whether the traveller moves from end to centre the wetted sector angle
— whether the irrigated areas are isolated or adjacent to each other

Generally, hose-drag travellers are operated for the full length of the tow-path in one direction. On the other hand, hose-pull travellers are usually operated from one end to the centre. By establishing the standing positions and times of the traveller, the actual length of hose can be determined. After that the inlet pressure required by the hose can be established.

Standing Positions

It is generally recommended that the ideal starting (initial) distance (SD) from the field edge should be 2/3R, where R is the radius of the wetted circle:

$$SD = \frac{2R}{3} \tag{4}$$

The final distance, (FD) from the other field boundary is:

$$FD = \frac{2}{3} \times \left[1 - \left(\frac{360 - w}{180}\right)\right] \times R \tag{5}$$

Where:

SD = initial starting point distance from the field edge (m)

FD = Final or stopping point distance to the end of the tow-path for end-to-end operation (m)

w = portion of circle receiving water (degrees)

R = wetted radius of sprinkler (m)

Assuming the farmer chooses option 3, the distance from the field boundary, to the starting standing position is:

$$SD = \frac{2 \times 69.34}{3} = 46.2 \text{ m}$$

If the sprinkler has a full circle coverage, then SD=FD. However if it is a part circle with say a 285 degree coverage, then the distance from the final standing position to the other field edge is:

$$FD = \frac{2}{3} \times \left(\frac{1 - 360 - 285}{180} \right) \times 69.34 = 27.0\text{m}$$

Having established the standing positions in terms of their distances to the 2 field edges, the standing time can now be calculated. The total standing time is equal to the time which the traveller should have taken to travel the rest of the distance had it not stopped. The general equation for standing time is:

$$t = \frac{2}{3} \times \frac{R}{V} \tag{6}$$

Full circle coverage standing time

For full circle coverage, the starting standing time (St) is equal to the final standing time (Ft) that is:

$$St = Ft = \frac{SD}{V} = \frac{2}{3} \times \frac{R}{V} \tag{7}$$

This gives the total standing time at each field edge as:

$$St = Ft = \frac{46.2}{0.362} = \frac{2}{3} \times \frac{69.34}{0.362} = 127.6 \text{ min utes}$$

or 2.13 hours for each field end. This means that the traveller irrigates while in motion for a distance of 307.6 m {400-(46.2 x 2)} Since the traveller speed is 0.362 m/min, the traveller irrigates while in motion for 14.16 hours. The total irrigation time is 18.42 hours (14.16 + 2.13 + 2.13).

Part circle coverage standing time

For part circle coverage, the individual standing times are equal to 127.6 minutes {46.2 m/(0.362 m/min.)} on one end and 74.6 minutes (27.0 m/(0.362 m/min.)} on the other. The average of these standing times is 101.1 minutes. The average standing time for part circle sprinklers can also be calculated using Equation 8 as follows:

$$St = Ft = \frac{(SD + FD)}{2 \times V} = \frac{2}{3} \times \frac{R \times W}{360 \times V} \quad (8)$$

This gives for a 285 degree part circle coverage:

$$St = Ft = \frac{(46.2 + 27.0)}{2 \times 0.362} = \frac{2}{3} \times \frac{69.34 \times 285}{360 \times 0.362} = 101.1 \text{ min utes}$$

In practice the two standing times are just averaged as assumed in the equation above.

Hose Length Determination

Having calculated the standing positions and times, the length of the hose can now be determined. The original assumption was that the hose should be as long as the distance from one end of the field to another, along the tow-path length of 400 m. However, it can be reduced by a minimum length corresponding to FD or in practice, by a maximum length corresponding to the average of SD and FD. In the case of the part circle coverage, the latter would be: ½(46.2 + 27.0) = 36.6 m, say 36 m. Therefore the actual length of the hose would be 364 m (400 - 36).

Head Loss in the Hose

Table 4 provides typical frictional loss gradients for the normal flow ranges of lay flat hoses operating at about 70 m head. This will be used as an example for frictional loss calculations. Manufactures' charts should be sourced when such calculations have to be done for different operating heads and hoses. Alternatively the general equation for computing friction losses in smooth plastic pipes and hoses, shown below, should be used.

Table 4 is initially used to estimate the friction losses for the lay flat pipes. Lay pipes collapse when they are drained. They are used with hose-drag travellers. Lay flat pipes tend to increase in diameter during operation and therefore can in fact take a higher discharge. Therefore there is a need to establish the diameter during operation. Once the inside diameter is determined, the equations for computing frictional loss gradients can be used to estimate more accurately the friction losses during operation. It has been established that the inside diameter of a lay-flat hose increases by about 10% under normal operating pressures thereby allowing the hose to be capable of delivering 20% more discharge than the same diameter rigid plastic hoses used in hose pull systems. The thick-walled plastic hoses used for hose-pull traveller have known inside diameters which can be used in the same equations for computational purposes. Their diameters do not change as a result of water pressure.

Table 4. Estimated friction loss gradient values in m per 100 m, for lay-flat irrigation hose operating at approximately 70.3 m pressure rating

Flow rate	*Nominal lay flat hose diameter (mm)*				
(l/s)	*63.5*	*76*	*102*	*114*	*125*
6.3	3.7				
9.5	7.9	3.2			
12.6	12.9	5.5			
15.8		8.3	2.2		
18.9			3.1	1.4	
25.2			5.3	2.9	
31.5			8.1	4.8	2.5
37.9			11.3	6.1	3.7
44.2				8.2	4.9
50.5				10.5	6.2
56.8					7.9
63.1					9.2

Having established that the length of the hose is 364 m, the friction loss through it can now be determined. From Table 4, the flow rate to 34.5 l/s is between 31.5 and 37.9 l/s, with Hf values of 4.8 and 6.1 m per 100 m for a 114 mm diameter pipe. By interpolation, we derive a Hf of 5.4 m per 100 m for the 34.5 l/s flow when using 114 mm lay flat hose. Therefore the total head loss is 19.66 m (5.4 x 364/100). Since the sprinkler operating pressure is 56.24 m and the total head loss of the hose is 19.66 m, the total operating pressure at the beginning of the hose will be 75.90 m, which exceeds 70.3 m pressure rating of the hose. Therefore, a bigger size hose will be selected, so that the total head loss is reduced to within 70.3 m. The next size up is 125 mm in diameter. By interpolation, the Hf is 3.06 m per 100 m.

Therefore the total head loss is 11.1 m (3.06 x 364/100). This gives a total operating pressure of 67.34 m (11.1 + 56.24). This suggests that we may be confronted with the same situation of exceeding the pressure rating of the lay flat hose, especially if we add the head losses in the riser. Under these circumstances, it is advisable to have the main line dissecting the plot in two, thus reducing substantially the head losses in the hose and possibly allowing the use of smaller diameter hoses. In order to establish the length of the short hose, it is necessary to go back to the calculation of standing time. In this case it is only at the edges of the field. As mentioned earlier, the distance from the field boundary is 46.2 m. Therefore the length of the hose will be 153.8 m (200-46.2).

This will reduce the head losses to 4.7 m (3.06 x 153.8/100). In selecting the pipe diameter, consideration should also be given to the cost and expected useful life of the hose as provided by the manufacturer, and whether the size of the traveller can drag the hose. This information is also provided by the manufacturer.

The general equation for calculating frictional losses in smooth plastic pipes and hoses is Equation 9, which is:

$$Hf_{100} = \frac{K \times \left(\frac{Q}{C}\right)^{1.852}}{D^{4.832}} \qquad (9)$$

Where:

Hf_{100} = friction losses (in metres) over a 100 m distance (m/100m)

K = constant 1.22×10^{12}, for metric units
Q = discharge in l/s
C = coefficient of retardation based on type of pipe materials (C = 140 for plastic)
D = inside diameter (mm)
Using the equation for a flow of 34.5 l/s and a 125 mm diameter hose gives:

$$Hf_{100} = \frac{1.22 \times 10^{12} \times \left(\frac{345}{140}\right)^{1.852}}{D^{4.832}} = 6.72 \qquad (10)$$

Therefore the head losses in the 364 m hose length are 24.46 m (6.72 x 364/100). This again suggests the use of shorter length of hoses.

Selection of the Traveller

Normally turbine drive travellers are used. The traveller will have frictional losses within it as well as the required drive turbine pressure at a given discharge.

Total Dynamic Head Requirements

Pressure requirements at mainline valve hydrant

The pressure requirements at the inlet of the system consist of the sprinkler operating pressure, friction losses in the hose, traveller head losses, head losses through the shut off valve and the height of the riser.

Table 5. Hose inlet pressure requirements

Type	*Head Loss (m)*
Sprinkler operating pressure	56.24
Friction loss in hose	4.7
Head loss in traveller (assumed)	4.7
Head loss in automatic valve	2.0
Riser height	2.5
Required pressure at inlet	70.14

If the pressure loss through the shut-off valve is assumed to be 2 m and the height of the gun riser is 2.5 m from the ground, then the inlet pressure would be calculated as shown in Table 5.

Head Loss in Mainline

The length of the water supply line from the river to the last hydrant is 588 m (630 - 52 +10), as the edge of the field is 10 m away from the pumping station and the furthest hydrant is one half of the tow-path width 52 m (1/2 x 104) away from the field edge. Therefore the total friction loss over the whole length of the supply line would be 5.88 x 1.6 m that is 9.4 m. This is within the 20% allowable pressure variability. The mainline does not necessarily have to be buried. It can also be portable.

Total Dynamic Head

The elevation difference from the source of water to the highest point is approximately 3 metres. Therefore the total dynamic head is 92.69 as shown in Table 6 below. The final design layout is shown in Figure 1.

Table 6. Total dynamic head requirements

Type	*Head Loss (m)*
Sprinkler operating pressure	56.24
Friction loss in hose	4.7
Head loss in traveller	4.7
Head loss in automatic valve	2.0
Riser height	2.5
Friction loss in mainline	9.4
Suction head	2.0
Total Head	81.54
10% for fittings	8.15
Elevation difference	3.0
Total dynamic head requirements	92.69

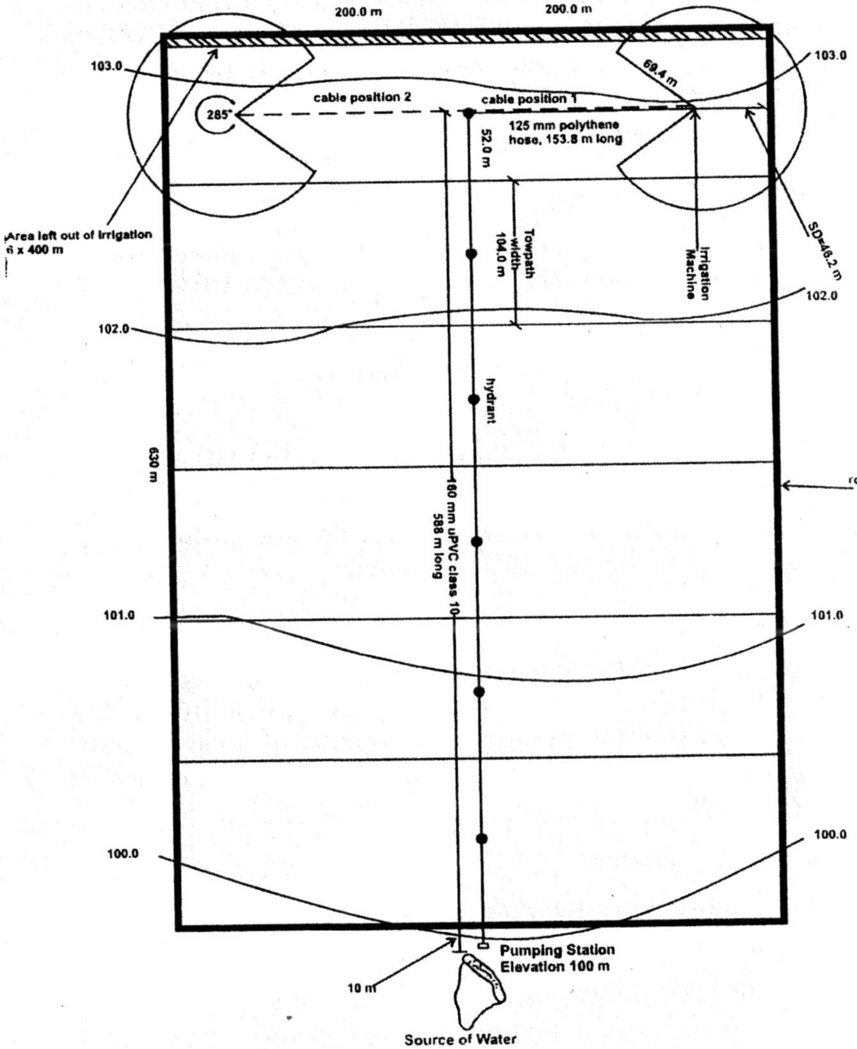

Figure 1. Final system layout for traveller irrigation design (field: 630 m x 400 m)

In comparison to the gun, if a boom with spray nozzles is used, the total head requirements can be reduced by 35 metres, since the total head requirements (friction losses in hose + traveller + nozzle operating pressure) of these systems do not exceed 30 metres. However, since the boom systems are smaller, a number of them operating at the same time would be required to cover the whole area.

Power Requirements

Through the application of Equations 9 and 10 presented, the power requirements are calculated as shown below:

$$\text{Power requirements} = \frac{34.5 \times 3.6 \times 92.69}{360 \times 0.6} = 53.3 \text{ kW}$$

or

$$\frac{34.5 \times 3.6 \times 92.69}{273 \times 0.6} = 70.3 \text{ HP}$$

If 20% is added for derating purposes, the prime mover would have to be 64.0 kW or 84.4HP.

System Components

Generally the major components of travelling irrigators are the irrigating unit, the supply pipe and a propelling mechanism (turbine pump). The example of a cable-drawn irrigator will be used to illustrate the major components in more detail.

The Irrigation Machine

The irrigation machine consists of:

— a trailer

— the pipe supplying water to the gun

— a winch and a hydraulic piston motor

The trailer is mounted on 4 wheels. The front wheels are able to turn and they form one piece with the towbar. The supply pipe has a hose coupler on one end and a gun on the other. The machine also carries the mechanism for stopping movement and irrigation once the machine reaches the other end.

The Hose Reel Trailer

The frame of the trailer is mounted on 2 wheels and carries the hose reel. The large horizontally mounted reel has a small cable-winding reel also mounted on top of it. The system also carries an air compressor and a mechanism for guiding and cleaning the hose as it winds.

The Hose

The hose is made in such a way as to be extremely wear resistant against the strain of traction and the high pressure it is subjected to. It is usually made of canvas-reinforced rubber. It has to be flexible.

The Gun sprinkler

Figure 2 shows a typical gun sprinkler.

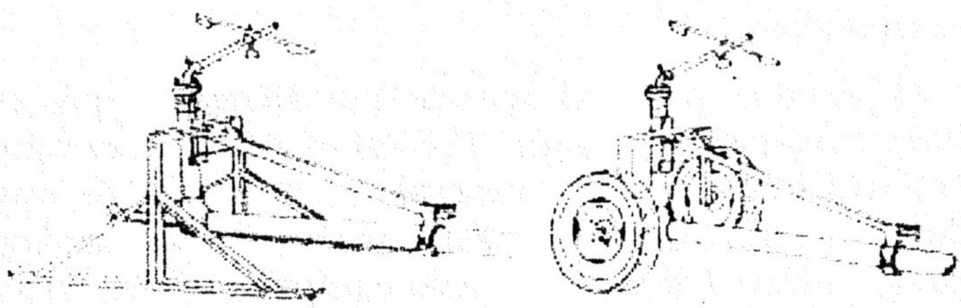

Gun sprinkler on skids *Gun sprinkler on wheels*

Figure 2. Typical gun sprinkler mounted on skids or wheels

Bill of Quantities

The source of water is 588 m away from the furthest hydrant. A 160 mm diameter, class 10 PVC pipe will be buried and it feeds the 6 hydrants along the mainline. The trench for the pipeline will be 1.00 m deep and 0.6 m wide. Manual labour will be used both for excavation and backfilling of the trenches. One access road will be constructed around the 400 m x 630 m field and outer drain will be provided. No new fencing is required since the farm is fenced already. The field toilets are already in place. It has also been assumed that other equipment such as tractors is already available for other farm uses.

The procedure for calculating labour requirements outlined as follows:

- setting out 588 m of pipeline and 2060 m of access road with drains, would require (2 648/1 600) 1.7 (say 2) workdays for the surveyor and (4 x 1.7) 6.8 (say 7) workdays for unskilled labour
- construction of 2060 m of road with drains would require approximately 10 (2060/200) machine hours
- excavation of a 1 metre deep trench 588m in length would require 147 (588/4) workdays for unskilled labour
- pipe laying would require 0.5 (588/1 000) workdays for skilled and (0.5 x 2) 1 workday for unskilled labour.

System Operation

There are several ways to operate the different types of travelling irrigation machines. The cable drawn travelling irrigator will be used as an example to illustrate the way travellers are operated. This system has got a trailer carrying the gun sprinkler. The trailer is also equipped with a water powered winch and a cable. The winch is driven by water pressure from the pumping unit. The gun sprinkler is supplied water from the pump, via a mainline, which has hydrants onto which the hose is connected.

The following procedure is typical for the way such a system is operated, step by step:

1. The tractor, hose reel and irrigating unit are harnessed in that order along the tow-path
2. The cable is anchored at one end of the field
3. The tractor is driven to the hydrant

4. The hose is then connected to the hydrant
5. Unwinding of the cable and hose are then done by driving the tractor to the other end of the field
6. The next step is to disconnect the hose
7. The hose reel and the irrigating unit are then brought back to the first position
8. The hose is attached to the irrigating unit and the unit is also detached from the hose reel
9. The hose reel trailer is then driven to the position where the cable is anchored
10. The pumping station should then be started

The irrigating unit will then start to operate. As it irrigates, it winds the cable on the winch and in the process pulls itself on the cable. Once it reaches the pre-determined distance close to the other end, it automatically stops moving and irrigating. If standing time is allowed for, it stops moving but continues to irrigate during the standing time. The following procedure should be followed when changing position to the next tow-path.

1. The hose is disconnected from the hydrant as well as from the irrigating unit
2. The hose is then connected to the hose reel
3. The tractor should be connected to the hose reel before draining the hose by air pressure from an air compressor
4. The hose is then rewound and the equipment is moved to the next tow-path

Maintenance of Travelling Irrigators

The maintenance of pipelines and pumping equipment of a traveller irrigation system is similar to that of the other sprinkler systems. The maintenance of the traveller irrigator requires skilled expertise, which should be provided by the dealer. There are a number of protection devices, which can be incorporated in the traveller system in order to improve the operation of the system.

Travellers irrigate for long periods unattended. Therefore there is a need to install system protection devices within the system for the shut down of the traveller system in the event of a malfunction. Similarly, the main pump supplying the irrigator should have pressure detectors in the main supply line so that when blockages or bursts occur the whole system can be turned off. Cable drawn (hose-drag) and hose reel (hose-pull) irrigators have an automatic pressure switch-off facility, which turns off the winding mechanism at the end of the run. It is not usual to cut off the main supply at the same time. However, as this can lead to excessive pressure build-up in the main, a cut-off mechanism can be provided for the motor of the pumping plant so that when the traveller cuts off, the motor can also turn off at the same time.

The irrigation machines should be moved in such a way as to avoid contact with power lines. Generally the machines should move parallel to power lines to minimise the contact of power lines and water. There has to be a minimum of 15 metres between the nozzle and power line to allow the water to break up into droplets before hitting power lines.

References

FAO, *Mechanized Sprinkler Irrigation, FAO Irrigation and Drainage Paper,* 1982.

Kay, M., *Sprinkler Irrigation: Systems and Practice,* Bedford, 1986.

Keller, J. and Bliesner, R.D., *Sprinkler and Trickle Irrigation,* Chapman and Hall, New York, 1990.

Withers and Vipond, *Irrigation Designs and Practices,* 1974.

5

Evaporation Loss during Sprinkler Irrigation

During sprinkler operation, water evaporates from droplets sprayed through the air. The amount of evaporation depends on three factors: (1) the climate demand; (2) the time available for evaporation to occur; and (3) the surface area of the water droplets.

Climate demand: The climate demand is a measure of the energy available for evaporation and the capacity of the air to store and transmit water vapour. The evaporation process requires 580 calories of energy to convert 1 gram of water from the liquid to the vapour form. This energy must be available from the environment surrounding the sprinkler. Energy is readily available during summer months, as indicated by high air temperatures and high levels of solar radiation. Conversely, low temperatures and low levels of solar radiation during winter months provide less energy for evaporation.

Air humidity is often expressed as relative humidity, which ranges from nearly 0 to 100%, with low values indicating dry air and high values indicating moist air. Since dry air has a greater capacity for moisture, evaporation will occur more rapidly when the air is dry than when it is moist. When the relative humidity is 100 percent, the air is saturated at its current temperature and cannot hold any more water; thus, no further evaporation occurs. This condition exists after dew-fall at night, during foggy conditions, and during rainfall. Also, since humidity levels are normally high during early morning and early evening hours in humid areas, evaporation rates are low during these times on most days.

Wind increases evaporation rates by transporting water vapour away from the evaporating surface. Wind also increases evaporation by transporting warmer or drier air from surrounding areas to displace the moist, cool air above an irrigated surface.

Time available for evaporation to occur: The time available for evaporation to occur is the relatively short time beginning when a water droplet leaves the nozzle and ending when it falls on the ground or vegetated surface. When water is sprayed at greater heights and over greater distances, the opportunity time for evaporation is increased. Thus, more evaporation would be expected to occur from a sprinkler installed on a tall riser than from the same sprinkler (operated at the same pressure so that the drop size distribution is the same) installed on a short riser.

Wind speeds are faster at greater heights above the ground surface, where there are few obstacles to air movement. Thus, evaporation loss from sprinklers mounted on tall risers is also increased because of these higher wind speeds.

Surface area of water droplets: Because evaporation occurs from the surface of the water droplets, the total surface area of the water droplets greatly affects the amount of evaporation loss. For a unit volume of water, the surface area doubles as the droplet diameter decreases by half. For this reason, evaporation rate increases as droplet size decreases if other factors remain constant, and the factors that cause droplet size to decrease will cause evaporation loss to increase.

Wind Drift Loss

Wind drift loss occurs when wind carries water droplets away from the irrigated area. Droplets may either evaporate while they are being transported or may fall outside the irrigated area. In either case, the water is lost because it is not available to the plants being irrigated. Wind drift loss increases as wind speeds increase and as droplet sizes decrease. At higher wind speeds, larger droplets can be transported by the wind, and droplets can be transported greater distances.

Sprinkler Evaporation Loss

Many studies of evaporation loss during sprinkler irrigation have been conducted. In their 1960 article, Frost and Schwalen summarized the results of 700 field tests conducted under a variety of climate conditions. They developed a nomograph that enables the user to estimate the percentage of evaporation loss during sprinkler irrigation as a function of sprinkler characteristics, operating pressure, and climate factors. They concluded that evaporation loss could be accurately estimated using three climate factors: air temperature, relative humidity, and wind speed. The Frost and Schwalen nomograph has been widely used to estimate sprinkler evaporation loss.

Seginer found strong correlations between sprinkler evaporation loss in Israel and both solar radiation and air temperature. He found weaker correlations with relative humidity and wind speed. He concluded that evaporation loss due to wind drift was negligible for the 4.5 to 6.0 km/h wind speed conditions.

Myers *et al.* conducted wind tunnel tests of sprinkler evaporation loss for typical Florida climate conditions. They stated that it is unlikely that evaporation from water droplets in transit from the sprinkler to the ground or plant surface could represent more than 5 % of the water applied by typical sprinkler irrigation systems under Florida climate conditions.

Evaporation Nomograph

The Frost and Schwalen nomograph can be used to estimate evaporation loss for specific climate and operating conditions. This nomograph allows the user to estimate the amount of evaporation loss for different times of day or for day-time and night-time irrigation.

The Frost and Schwalen nomograph consists of vertical columns numbered 1 through 9. Data must be entered in columns 1, 2, 5, 7, and 9. The nomograph is used by drawing straight lines that connect the input data to calculate intermediate factors in columns 3, 4, and 8. Finally, a straight line is drawn between columns 4 and 8 to provide an estimate of the percentage of evaporation loss in column 6.

The use of the nomograph is illustrated by the following example: Estimate the percentage of evaporation loss from a sprinkler with a 3/16-inch nozzle diameter operated at 40 pounds per square inch (psi) when wind speed is 5 miles per hour (mph), temperature is 90°F, and relative humidity is 10 percent. The sprinkler discharge rate in this example is 6.3 gallons per minute (gpm).

The calculations are done in the following steps:

1. Draw a straight line from the point representing 10 percent relative humidity in column 1 through the mark for 90°F temperature in column 2. Extend this line to indicate a vapour-pressure deficit of 0.63 psi in column 3.
2. Draw a straight line from the point representing 0.63 psi vapour-pressure deficit in column 3 to the mark that indicates 3/16-inch (12/64-inch) nozzle diameter in column 5. Locate point A where this line intersects the pivot line in column 4. Point A will be used in step 4 below.
3. Draw a straight line from the mark for 40 psi nozzle pressure in column 7 to the point representing 5 mph wind speed in column 9. Locate point B where this line intersects the pivot line in column 8. Point B will be used in step 4 below.
4. Draw a straight line from point A on the pivot line in column 4 to point B on the pivot line in column 8. Read the percentage of evaporation loss where this line

intersects column 6. For this example, evaporation loss is 8.5 percent.

Thus, 8.5 percent of the water discharged from an irrigation system using 3/16-inch nozzle diameters and operating at 40 psi would evaporate under these extreme climate conditions. For the 6.3 gpm sprinkler discharge rate, 8.5 percent, or 0.54 gpm, would be expected to evaporate, while the remaining 5.76 gpm would be expected to reach the soil surface or plant canopy.

The combined effects of temperature and relative humidity determine the vapour pressure deficit. The vapour pressure deficit is a measure of the dryness of the air; thus, it is an index of the evaporation rate. Figure 1 shows the general relationships among air temperature, relative humidity, and vapour pressure deficit from the Frost and Schwalen nomograph. This figure illustrates that the vapour pressure deficit (and evaporation rate) will be highest when temperatures are high and the relative humidity is low. Notice that when temperatures are low, the vapour pressure deficit will be low, despite the humidity level. Also, when the relative humidity is high, the vapour pressure deficit will be low, despite the temperature.

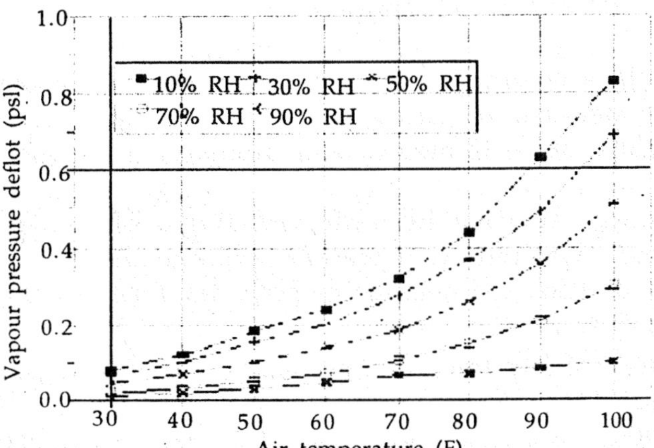

Figure 1.

Because high humidity levels are typical, the vapour pressure deficit is typically low despite the high temperatures. Figure 2 shows monthly vapour pressure deficits. Both maximum and average monthly values were calculated. The maximum values were calculated from the maximum temperature and minimum relative humidity data.

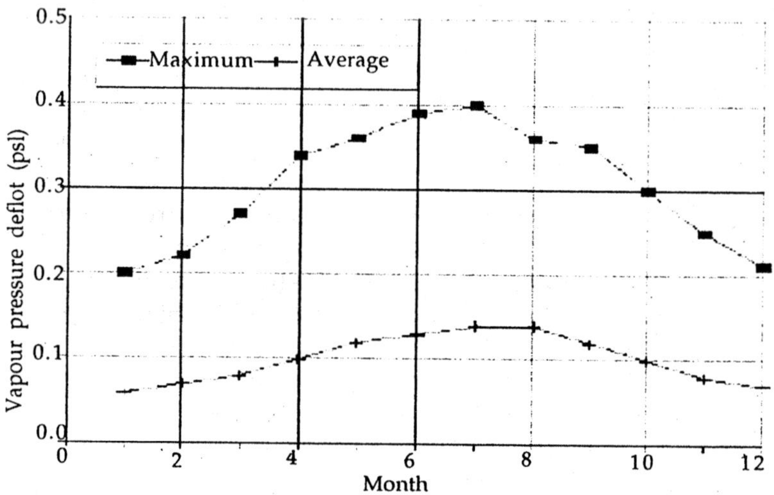

Figure 2.

These values would be expected to represent peak vapour pressure deficits occurring during the early afternoon, when temperature is highest and humidity is lowest. The average values were calculated from average temperatures and average relative humidity readings. These values would be expected to represent average daily vapour pressure deficits, the average of both day-time and night-time conditions.

Figure 3 shows the monthly distribution of average and maximum daily evaporation percentages. The average daily values were low and uniform at about 1.5 percent throughout the year. The maximum (early afternoon) values were more variable, ranging from 2.1 percent in January and December to 3.1 percent in July.

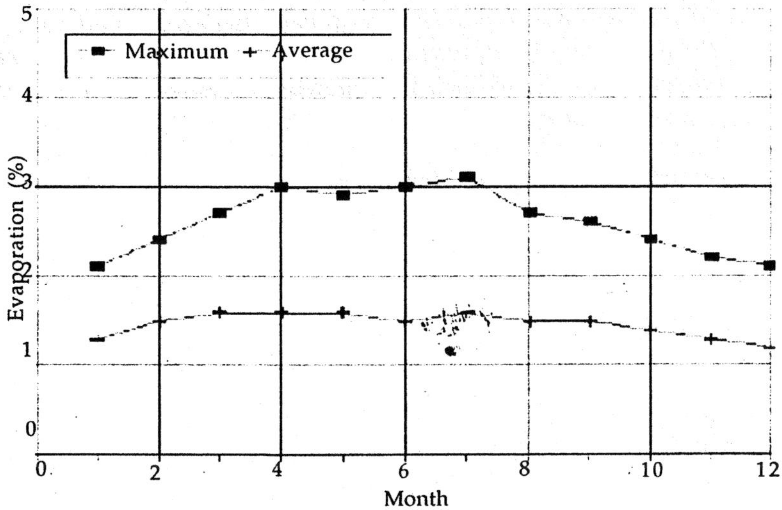

Figure 3.

Reducing Evaporation Loss from Sprinklers

Evaporation loss from a sprinkler irrigation system can be reduced by changing the sprinkler operating conditions to increase the sizes of water droplets or by irrigating when climate demands are low. Evaporation loss will be reduced by changing either or both of the following operating conditions: (1) increasing the nozzle diameter, or (2) decreasing the operating pressure. Both of these changes will increase the proportion of large droplets, thus decreasing the surface area from which evaporation can occur. However, sprinkler users must follow the manufacturer's recommendations when selecting nozzle diameters and operating pressures. Otherwise, poor distribution patterns will occur, reducing the uniformity of water application.

The effect of changing the nozzle diameter is illustrated in Figure 4. As shown in Figure 4, the evaporation percentage is large when small nozzles are used. It is small when large nozzles are used and other factors remain constant. Notice that the evaporation percentage is not linearly related to the nozzle diameter. It increases rapidly when nozzle sizes are decreased below 5/16-inch but decreases slowly when nozzle sizes are increased.

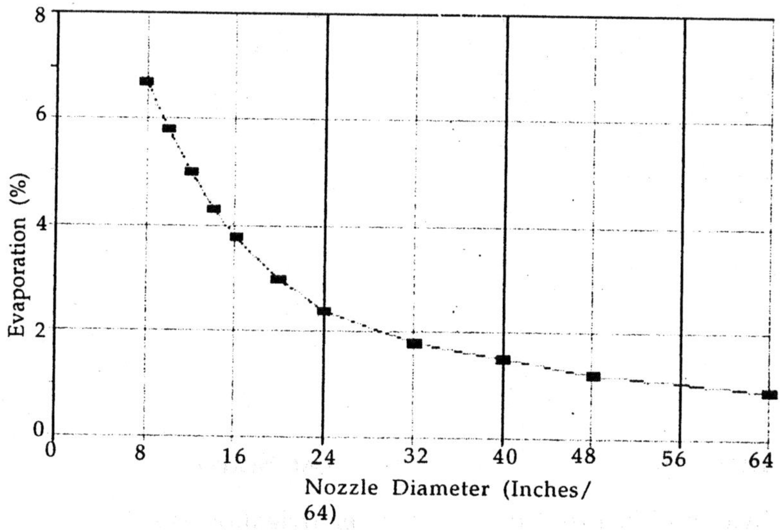

Figure 4.

The effect of changing the operating pressure is shown in Figure 5. Calculations for this example were made using 5/16-inch nozzle diameters. As Figure 5 illustrates, the evaporation percentage is small when small pressures are used; it is large when large pressures are used and other factors remain constant. Also, the relationship between evaporation percentage and pressure is approximately linear. It increased from 1.4 percent at 20 psi to 5 percent at 80 psi.

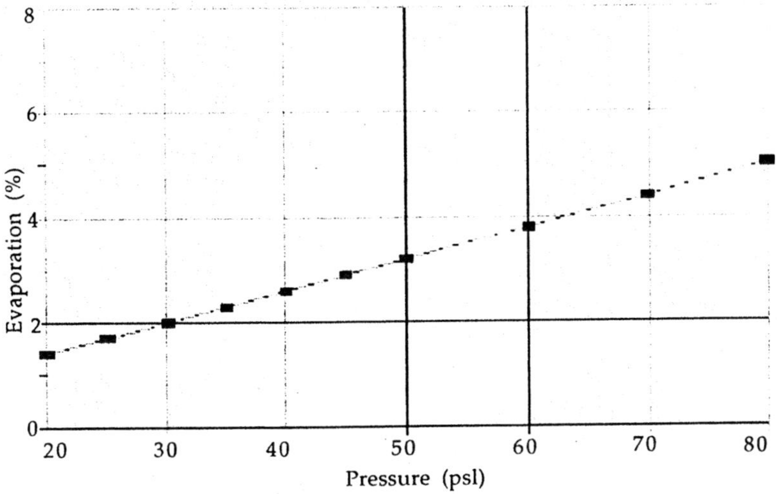

Figure 5.

The use of excessively large nozzles or excessively small pressures must be avoided. Excessively large nozzles will increase sprinkler and distribution pipeline flow rates, making the system cost too high. Also, soil infiltration rates may be exceeded, causing runoff. Thus, allowable increases in nozzle sizes will be limited by economic considerations and production system characteristics.

Excessively low pressures will reduce the uniformity of water application and waste water because of low application efficiencies. Sprinkler mechanisms will not operate properly at pressures that are too low. Excessively low pressures also will produce droplet sizes that are too large, creating the potential for crop damage or soil erosion from the impact of the droplets. Thus, allowable reductions in pressures will be limited by sprinkler operating characteristics and the requirement for uniform water applications.

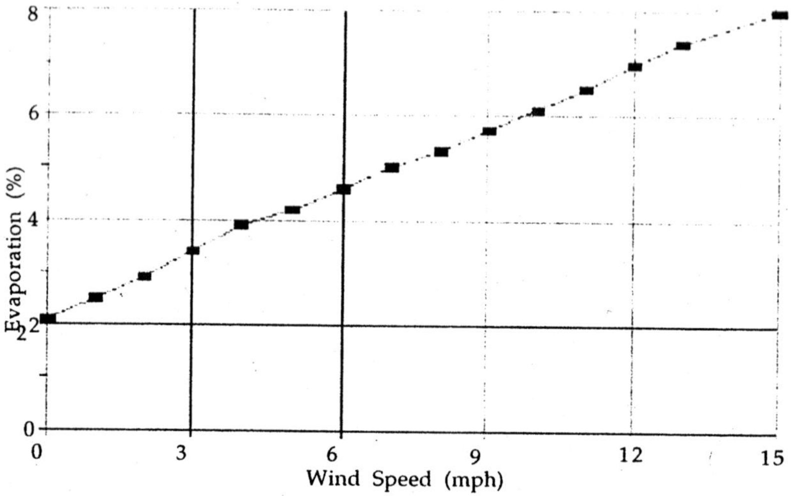

Figure 6.

Evaporation loss can be reduced by operating sprinklers only when climate demand is low. This means operating sprinklers when relative humidity is high and air temperature and wind speed are low. The relative humidity is very high during night-time, early morning, and late evening hours on most days. At those times, air temperatures are also low and wind speeds are normally reduced, minimizing sprinkler evaporation loss. Figure 6 illustrates the effects of wind speed on evaporation loss with 5/16-inch nozzle diameters, and operating pressures of 45 psi. For this example, the relationship between evaporation percentage and wind speed ranging from 0 to 15 mph was approximately linear. Evaporation loss increased from 2.1 percent to 8.0 percent as wind speed increased from 0 to 15 mph.

Reducing Wind Drift Loss

Wind drift loss occurs when small water droplets are transported from the irrigated area by wind. This occurs when droplets are small and wind speed is high.

If the wind is blowing when irrigation occurs, some wind drift loss is unavoidable. To avoid excessive wind drift loss, irrigation systems should not be operated when wind speeds are high. Also, a sprinkler should be operated only within the operating pressure range for which it was designed. If operating pressures are too high, too many small droplets will be created and wind drift loss will increase. Most sprinkler companies manufacture nozzles specifically designed to reduce the number of small droplets and minimize wind drift loss.

Increasing Evaporation Loss from Sprinklers

Under some conditions, it may be desirable to increase the amount of evaporation loss from sprinklers. Such increases would be desirable when sprinklers are being used for crop cooling, as is the case during the establishment of transplants, or when water disposal is the objective—for example, in the design of spray-fields for land applications of waste-water.

Thus, sprinkler irrigation is used to prevent excessive water stress while plant roots (normally, those of young plants or transplants) are developing. It is often required to establish such fruits and vegetables as strawberries and cabbage, and many ornamental plants.

In the design of spray-fields for waste-water disposal, the size of the required spray-field can be reduced if evaporation loss is increased. Such reductions are possible because some of the water discharged from sprinklers evaporates before it reaches the ground or plant surface.

To increase evaporation loss, small nozzles should be used and sprinklers operated at high pressures and when climate demand is greatest. Climate demand will normally be greatest during the early afternoon, when relative humidity is lowest and air temperature and wind speed are highest.

Excessively high operating pressures should be avoided because both diameter of coverage and water application uniformity will be reduced if the pressure is too high. Also, sprinkler mechanisms may not operate properly at excessively high pressures.

Under windy conditions, the distribution of the very small water droplets created by excessive pressures will be easily distorted by wind. These small droplets may be transported to surrounding areas by wind drift. Under calm conditions, these small droplets will fall nearer the sprinkler, reducing the effective diameter of coverage and possibly reducing the uniformity of water application. Thus, the maximum operating pressure should be limited based on the sprinkler manufacturer's recommendations so that acceptable diameters of coverage and uniformity of water application may be achieved.

Spray Evaporation Losses from Sprinkler Irrigation Systems: an Experimental Study in Southern Manitoba

Interest in water conservation is increasing as a result of ever increasing demands on scarce water resources. Properly designed irrigation systems can minimize the losses of the water delivered to the plants. Among the different systems used for irrigation, the sprinkler irrigation system is one of the popular methods for achieving high application efficiencies. In Manitoba, most of the field crop irrigation is done by sprinkler irrigation systems consisting of travelling guns, side-roll systems, linear move systems, or centre pivot systems. In a sprinkler irrigation system, very little water is lost in the conveyance system up to the sprinkler nozzle. Therefore, most of the losses are expected to take place from the moment the water leaves the nozzle until it reaches the soil within the root zone.

The various components of the losses in irrigation systems need to be identified before any improvements on the design of irrigation systems can be made. As mentioned previously, the losses that occur up to the sprinkler nozzle are relatively minimal and are preventable. Once the water leaves the sprinkler nozzle, it travels through the air and reaches the crop canopy or the soil surface. The portion of the water that is lost by evaporation during the time it travels through the air and before it reaches the crop canopy is called "above canopy spray evaporation losses" (ACSEL). Another portion is intercepted by the crop canopy and part of this is evaporated back to the atmosphere.

Any drift losses that occur as a result of the wind are not considered a loss as long as the spray drift falls within the boundaries of the cultivated area. The drift component only affects the uniformity of water application. However, if the drift component leads to local over application of water which results in deep percolation below the root zone, it should be taken into account in calculating the overall farm irrigation efficiency. The water application efficiency is defined as the water beneficially stored in the plant root zone as a percent of the total amount of water delivered to the field. In sprinkler irrigation systems that are well-maintained, a major portion of the loss in the field occurs from the time water leaves the sprinkler nozzle until it reaches the root zone.

Losses at the field level include evaporation from spray droplets that are travelling through the air (ACSEL), the portion intercepted by the crop canopy, and water deep percolating below the root zone. The ACSEL does not include the portion of the water that is intercepted by the crop canopy. In a sprinkler irrigation system, ACSEL comprises a major component of the total losses occurring between the sprinkler nozzle and the plant root zone. This study was initiated by the Association of Irrigators in Manitoba to measure the ACSEL in southern Manitoba. The results from this study will help producers understand how much of the irrigation water is lost by spray evaporation under conditions prevailing in Manitoba.

There are many factors that contribute to spray evaporation losses including weather variables and types of equipment. This study follows two phases. In Phase 1, the equipment-related factors were identified in preparation for Phase 2. In Phase 2, the impact of weather-variables and other factors will be studied. The results from Phase I which determine equipment-related factors that influence spray evaporation losses are presented herein. The results show the ACSEL as a function of relative distance from the sprinkler nozzle.

The Phase 2 study to identify the impact of different weather variables will be repeated under different weather conditions. One of the assumptions is that the weather variables do not change during the period of testing which lasted about an hour. In Phase 1, the number of water samples collected per test ranged between 40 and 60. The smaller the number of water samples which must be collected, the shorter the duration of each weather-condition test. Therefore, any reduction in the number of samples that are still representative of typical losses from the nozzles, as determined in Phase 1, will be useful in the Phase 2 experiments that investigate weather variables. Phase 1 results will also help determine the number and proper placement of collectors for sampling water from different types of irrigation systems.

For the purpose of this study, the above canopy spray evaporation losses (ACSEL) were defined as the amount of water that evaporates while the water droplets travel between the spray nozzle and the crop canopy, calculated as a percentage of water that leaves the nozzle. Traditionally, the losses are determined from volumetric or gravimetric measurements of water collected in catch-cans. An inherent drawback in these volume-based, catch-can studies is that the spray evaporation losses are included with the wind drift losses. For the same nozzle height, droplet size, and exit pressure, wind drift losses are affected only by the wind speed and direction, while the spray evaporation losses are affected by both wind speed and other weather variables such as ambient temperature, vapour pressure deficit, relative humidity, etc. Since this project measured the ACSEL only, it was desirable to separate the wind drift loss component from the spray evaporation loss component. George used the electrical conductivity (EC) method to measure only the ACSEL component.

The electrical conductivity method is based on the premise that any loss or gain of water by the droplet travelling through the air will lead to a corresponding change in its solute concentration, and hence its electrical conductivity. As a droplet of water travels through the air, it loses water through evaporation. Because the initial solute concentration is usually very low in water used for irrigation, it can reasonably be assumed that there will be no precipitates out of the droplet as it is evaporating. As the water evaporates from the droplets the solute concentration increases. By measuring the changes in the electrolyte concentration of the water, it is possible to quantify the collective volume of water lost by the droplets as they travel through the air. As the water evaporates, the solute concentration increases, and the electrical conductivity also increases.

By measuring the EC of the source water and the EC of the water caught in individual collectors positioned just above the crop canopy, the spray evaporation loss was calculated using:

$$\text{Loss}(\%) = \frac{EC_c - EC_s}{EC_c} 100 \tag{1}$$

where

EC_C = electrical conductivity of water in the collector, and

EC_S = electrical conductivity of the source water.

In similar studies using the electrical conductivity method, Yazar reported the losses to be between 1.5 and 16.8% of the total sprinkled volume on tests performed in Nebraska. Yazar also found that wind velocity and vapour pressure deficit were the most significant factors affecting the losses.

George used the E_C method to measure the evaporation losses and found maxima near the sprinkler and at the periphery of the application diameter. These results are attributed to the fact that the spray droplets landing closest to the nozzle are the smallest. The droplets travelling to the outerside of the grid travel the greatest distance in the air and, therefore, evaporate more.

Frost and Schwalen found spray losses as high as 45% under extreme conditions of bright sunlight, high temperatures, and low humidity prevailing in Arizona. While these results were obtained from an individual sprinkler, the spray losses on the fringe area of a single sprinkler were not comparable with those occurring under field conditions with overlapping area between sprinklers. Losses were found by subtracting the amount of water arriving at the field surface from the amount of water being applied to the field. Till measured the spray evaporation losses using the change in concentration of chloride ions in the irrigation water travelling from the sprinkler nozzle to the ground. The results from this experiment were compared with that predicted using Frost and Schwalen's nomograph and were found to be in close agreement.

Seginer compared the consumptive use of water from a sprinkler irrigated crop with that from a crop irrigated with a near-the-ground irrigation system. The near-the-ground system was used as a control in which the spray losses and the interception losses were almost completely eliminated. The results indicated a marginally significant ($p = 0.1$) increase in consumption of about 10% for the sprinkler-irrigated crop that was irrigated every three days. However, no significant difference between the treatments was found for the treatments in which the irrigation interval was two weeks.

Using the EC method, Hermsmeier reported that evaporation from sprinklers could range from 0 to 50% over short periods. Hermsmeir also noted that evaporation during daytime hours is 3 to 4 times the evaporation during night-time hours during July and August in the Imperial Valley, California. The air temperature and the rate of application were found to be better factors for estimating sprinkler evaporation than wind speed or relative humidity. Hermsmeir found that pan evaporation data along with rate of application could provide a good estimate of sprinkler evaporation.

Spurgeon et al. reported that hot, dry, and windy conditions could cause spray evaporation losses from sprinkler irrigation systems to approach 30% of the water applied. In a volume-based study using rain-gauges conducted in Kansas by Steiner et al., the average spray loss was found to be about 15% under highly evaporative conditions. They also estimated the seasonal losses of plant-intercepted water to be about 2 to 4%.

There are many equipment-related factors such as droplet size, nozzle angle, and pressure that contribute to spray evaporation losses. Previous studies have shown that water droplet size is an important consideration in the design of sprinkler irrigation systems. Small droplets are more susceptible to wind drift and evaporation. Droplet size can affect both the wind distortion of the spray pattern and spray evaporation losses.

Equipment variables that affect the droplet diameter are the nozzle size, geometry, and operating pressure. Kohl and Wright and Dadiao and Wallender have shown that the size of the spray droplets was proportional to the nozzle diameter. In addition, many researchers have found that the spray droplet size at any distance from the sprinkler is related to the nozzle size. Dadiao and Wallender also found that the droplet size was a function of nozzle pressure. As the pressure increases, the initial velocity of the droplets increases leading to an increase in the distance travelled by the droplets. Many researchers have reported that the diameter of the nozzle played a major role in the break up of the droplets and indirectly influenced the evaporation losses. Frost and Schwalen found that a 25% increase of nozzle operating pressure increased the evaporation losses by 25%. They also noted that smaller nozzle diameters tended to break up the droplets leading to greater evaporation losses.

Chaya and Hills reported that for a given nozzle size, the droplet size was found to be inversely proportional to the operating pressure. In addition, the droplet diameter was also found to be proportional to the Reynolds number of the water. They also evaluated the sprayer design and found the fixedcoarse plate sprayers tended to produce more uniform droplet size than other sprayer designs. Spinners that had double jets leaving the orifice, produced smaller droplets compared to spinners with single jets. The droplet formation from jet breakups in microsprayers was found to be similar to that of conventional agricultural sprinklers. Vories and von Bernuth found that a sprinkler of a given nozzle size, trajectory angle, and operating pressure produced a set range of drop sizes.

Kohl and DeBoer observed that for low pressure spray type agricultural sprinklers, the geometry of the spray plate surface, rather than the nozzle size and operating pressure, was the dominant variable that influenced drop size distribution. They also found that smooth spray plates produced smaller droplets compared to coarse-grooved spray plates. Low pressure spray heads used in centre pivot and linear move systems tended to have similar components, which consisted of a nozzle that released a vertical jet of water onto a spray plate or oscillating-grooved cone rotating in a plane perpendicular to the water jet.

The low pressure heads operate at 100 to 175 kPa and the wetted diameter of the throw is about 10 m (compared to 30 m for impact heads). The low-pressure heads had a droplet size ranging from 1 to 3 mm compared to 3 to 6 mm diameter for impact-sprinkler heads. Low-pressure heads are usually mounted on retractable drop-tubes that bring the nozzles closer to the crop canopy.

Most of the above findings attributed the spray evaporation losses to weather variables and to some extent on the type of equipment used for testing. The applicable weather variables include: wind speed, vapour pressure deficit, air temperature, and solar radiation. Thus, any experiment to investigate the weather variables should ideally be conducted under controlled conditions with the ability to change each weather variable independently. Such an experimental set up is too expensive to implement. An alternative plan is to use irrigation systems under widely varying weather conditions with the hope that many different combinations of weather-variables could be captured over several seasons. This will require a large number of tests during which the weather conditions could reasonably be assumed to be constant during each test.

If the electrical conductivity method is to be used for such a study, the number and placement of collectors used in sample collection will determine the time required for each test. In addition, the placement of the collectors should give a representative evaporation loss. In this study, the spray evaporation loss data are presented as a function of distance from the nozzle for different types of irrigation systems. The following questions will be answered by these experiments. Is there a difference in spray evaporation losses from a single sprinkler system depending on where the sample is collected? Does the wind direction in relation to the spray pattern of the irrigation system affect the spray evaporation losses? Answers to these questions will also help plan the optimum placement of collectors for the determination of spray evaporation losses.

Materials and Methods

Table I shows a listing of the type of equipment tested in this study. The wind direction in relation to the average travel direction is reported for the travelling guns. The travelling guns throw water much farther compared to other types of systems and, therefore, are prone to wind drift. The actual sample sizes indicate the number of collectors used during each test except for the travelling guns. The samples collected from the extreme ends of the rows were not included in the sample size.

Up to 60 water samples had to be collected at various distances from the sprinkler nozzles. The collectors were cut from the inverted neck portion of two-litre clear plastic bottles (Fig. 7). The height of the container was 150 mm from the base to the top. Sixty catch-container stands were placed in the field in a grid formation. The stands were spaced at various distances depending on the wind velocity. Approximately four source water samples were taken for each test at 15 min intervals. The source water samples were collected directly from the water leaving the nozzles. Some systems had a diverter valve in the irrigation pipe supplying the water which could be used to collect the water sample. A weather station, consisting of a psychrometer, a wind direction indicator, and an anemometer was set up away from the equipment and from any trees or buildings. Weather data were recorded at 15 min intervals.

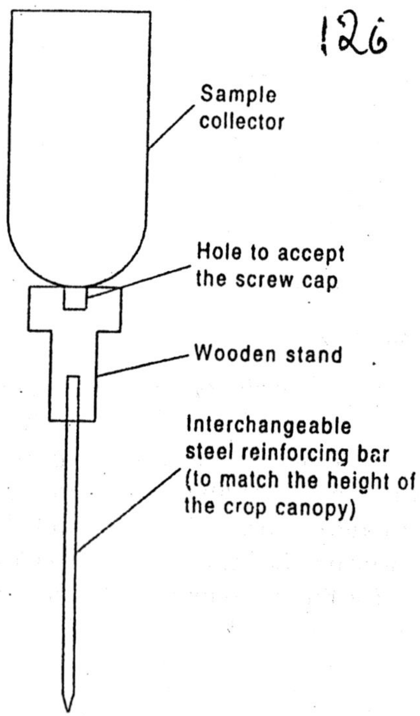

Figure 7. Schematic of the spray water sample collector.

Table I. Coefficient of variation and required sample sizes for the different tests.

Type of irrigation equipment		Actual sample size/row	Row 1		Row 2		Row 3		Row 4	
			CV	RSS*	CV	RSS	CV	RSS	CV	RSS
Travelling gun (ABI Irrigation AT110)• (Figs. 2 - 4)	0	16‡	0.13	8	0.10	4	n/a	n/a	n/a	n/a
	45	13‡	0.18	25	0.19	28	0.17	23	n/a	n/a
	90	10‡	0.09	7	0.09	5	0.08	5	n/a	n/a
Lockwood 2265 centre pivot with impact sprinklers (Fig. 5)	Test 1	15	0.31	1	0.22	1	0.90	20	0.20	1
	Test 2	15	0.56	2	0.45	1	0.33	1	0.27	1
	Test 3	15	0.34	5	0.32	5	0.15	2	0.25	2
Valmont 4271 centre pivot with flood jets (Fig. 7)	Test 1	15	0.31	7	0.21	2	0.35	7	0.31	6
	Test 2	15	0.69	3	0.58	4	0.61	3	0.62	4
Zimmatic Type 7 Model 510 centre pivot with flood jets (Fig. 8)										

Contd....

Contd....

	Test 1	15	0.17	3	0.38	10	0.21	5	0.16	3
	Test 2	15	0.12	2	0.16	3	0.22	6	0.29	13
Reinke Electrogator 60 KT centre pivot with two-way impact sprinklers (Fig. 6)		16	0.57	3	0.44	7	0.22	2	0.49	2

* Required sample size (RSS) is calculated based on = 0.05 and a tolerable evaporation loss precision of ± 0.5.

• Angle between wind direction and average travel direction of the travelling gun

‡ Sample size shown does not include the samples from collectors placed at extreme ends.

The collectors were positioned perpendicular to the direction of travel of the sprinkler irrigation system. Once the system passed over the grid, water samples were collected in 60-mL bottles from the collectors taking care to minimize any subsequent evaporation. It took approximately 30s to collect each water sample. The bottles were rinsed with the collected water once prior to any sample collection. This was done to ensure that any subsequent adsorption of the ions to the interior wall of the sampling bottle would be minimal.

The water samples were placed in the laboratory overnight to equilibrate with the room temperature. Rubber gloves were worn at all times when taking E_C readings to prevent contamination of the conductivity cell or the sample bottle. The electrical conductivity was measured for each sample using a YSI Model 32 Conductivity Meter. Only 1 mL of a water sample was required for the E_C measurement. The meter is capable of reading to µS with a precision of ±0.2% full scale. After the electrical conductivity measurements were completed, the collectors were placed in a sink filled with soapy water, washed, and then rinsed with hot water. The washed containers were finally rinsed with distilled-deionized water to remove any remaining solute adsorbed on the walls. The collectors were then hung upside down on drying stands by attaching the caps to elastic bands on the stands.

The containers were allowed to dry overnight and then placed in clean plastic bags for transportation to the field. The sample bottles were also washed with hot soapy water and rinsed with distilled-deionized water. Distilled-deionized water was poured into a few collectors and sample bottles and the electrical conductivity was determined to ensure the sampling equipment was relatively free of any contaminants. If the collectors came in contact with any soil particles, a potential for adding ions to the water samples occurred. Therefore, care was taken to avoid any contamination of the collectors during installation in the field.

The loss of water by evaporation from the collectors during the duration of the sample collection period was considered. Previous research has suggested the use of diesel fuel as an evaporation suppressant in the collectors. However, preliminary experiments done with the addition of the diesel fuel in the collected water caused the electrical conductivity to change as a result of contaminants in the diesel fuel. Even though the diesel fuel could be used as a suppressant in volume-based studies, it proved to be unsuitable for electrical conductivity-based studies. Therefore, another alternative method had to be developed to account for the evaporation during sample collection.

The water samples from the collectors were collected at random within each row of the grid. The time of sample collection was recorded. The electrical conductivities of the samples were plotted as a function of time and any observed trends in the data were analysed. If the spray loss of the sample changed with time, the spray loss as a function of time was plotted and an equation describing the best-fit-line obtained. This equation was used to adjust the measured spray losses based on the time of sample collection.

The first series of tests were carried out using an ABI Irrigation Model AT110 travelling gun irrigation system with a Big Gun SR50RN 24E nozzle. The grid was arranged such that the entire radius-of-throw of the gun was covered. Thus, the distribution of the losses as a function of distance from the nozzle could be determined. The tests were carried out only when a steady wind in the desirable direction was observed. Since all of the tests were carried out in the field, multiple rows of catch-containers were used to compensate for the inability to reproduce the wind velocities in the desirable orientations for a repeat test. The individual rows served as replicates for each test.

The coefficient of variation (CV) and the minimum sample size (N) was calculated for data from each row of collectors. The minimum sample size (N) was calculated using:

$$N = \frac{t^2 S^2}{E^2} \tag{2}$$

where:

t = 95% confidence limit for (n-1) degrees of freedom (n = number of samples/row),

s = sample standard deviation, and

E = tolerable precision in spray evaporation losses (± 0.5 %).

The precision of the conductivity meter is ± 0.2% of full scale in the range of electrical conductivities measured with the field samples. A larger tolerable precision for the spray evaporation loss was selected to account for other sources of variation during the test.

Results

Figures 8 through 10 show the spray evaporation losses as a function of distance from the gun. Since the tests were carried out at different times, the weather variables including the wind speed were different for the different tests. Therefore, the losses should not be compared between the different tests and types of equipment.

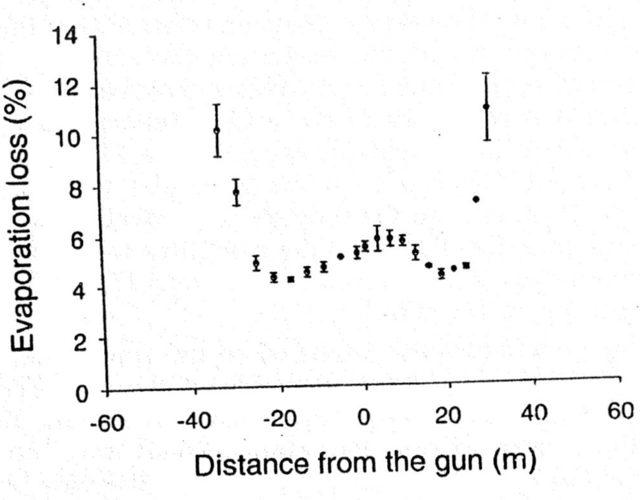

Figure 8. Travelling gun with wind direction at 0° to the average travel direction.

Results from three different tests shown in Figs. 6 through 4 were carried out for different wind speeds and directions, i.e. wind direction parallel, at a 45E angle, and perpendicular to the average travel direction of the gun. For all the tests, the gun rotated through an angle of 180E or 90E on each side of the line of travel. Thus, the negative distance from the gun refers to the sample collectors that were placed perpendicular to the centre line of gun travel but on the side where the nozzle approaches the line. Positive distances refer to locations on the side where the nozzle retreats from the centre line of gun travel. The error bars represent standard errors calculated using data from the different rows of collectors that were used as replications. Two to three rows of collectors were used as replicates for the tests carried out using the travelling guns. Table I lists the number of samples per row. The weather data was collected at about 15 min intervals to monitor any changes in conditions. Some tests had to be abandoned due to rain or gusty winds during the course of the tests.

Figure 8 shows the results when the wind direction was parallel to the average travel direction of the gun. In this case, the nozzle is throwing the water trajectory with the wind at the mid point. The average losses on either side of the nozzle vary between 4 and 6%, excluding the samples collected beyond 25 m from the nozzle. This is reasonable considering the fact that each side of the grid is exposed to the same environment. The maximum losses (11 to 12%) occurred at the greatest distance from the gun. The water collected beyond 25 m is from the droplets that travelled the greatest distance and, therefore, had the most time to evaporate. The minimum loss (4.3%) occurred at between 15 and 22 m from the gun on either side.

Figure 9 shows the results when the wind direction was 45E to the average travel direction of the gun. The water droplet trajectory is altered by the change in wind direction. On the windward side, the droplets could travel no further than about 30 m from the nozzle. On the other side, the wind seems to carry the droplets as far as 50 m from the nozzle. The evaporation loss on the windward side is about 2%

higher than on the opposite side. Since all other weather variables are the same, the difference could be attributed to the difference in wind velocity relative to the droplet velocity. The droplets in the windward side travel across the wind direction, while on the opposite side the droplets travel with the wind. The droplets that travel with the wind may experience a lower wind velocity compared to the droplets on the windward side. The characteristic increase towards the end of the trajectory is due to the differences in droplet size. For the travelling gun system, the CV presented in Table I was calculated after excluding data from the extreme ends showing high evaporation losses.

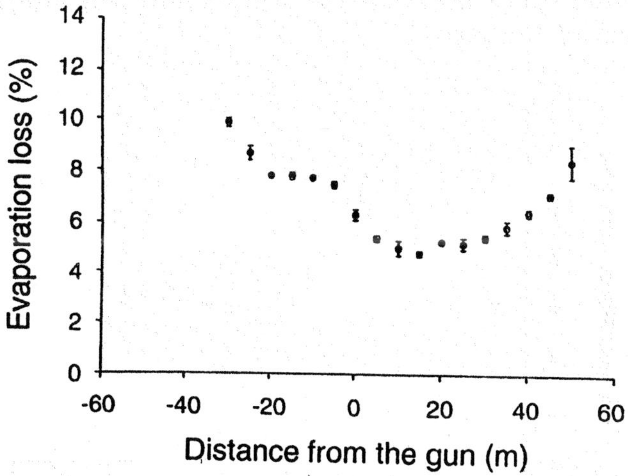

Figure 9. Travelling gun with wind direction at 45° to the average travel direction.

The relative wind velocity effect is even more pronounced in the test carried out when the wind direction was 90E to the average travel direction (Fig. 10). In this case, the water trajectory on the windward side becomes subjected to a higher relative wind velocity compared to the trajectory travelling with the wind direction. Once again, the higher losses towards the ends of the trajectories are due to the differences in the size of the droplets. The loss towards the end on the windward side is mixed with smaller droplets and shows a lower loss relative to the leeward side.

Figure 11 presents data obtained using a Lockwood 2265 Centre Pivot system equipped with Senninger 6E 5006 impact sprinklers located 4 m above the ground at spacings of 2.7 m. Four rows of collectors were used as replicates for the centre pivot systems. Tests 1, 2, and 3 result from three different tests carried out at different times under different weather conditions. Each test was carried out using four rows of collectors as replicates. Collectors were positioned in a direction perpendicular to the direction of travel of the sprinklers. Thus, the rows were placed parallel to the irrigation pipe between the towers towards the middle portion of the pipe.

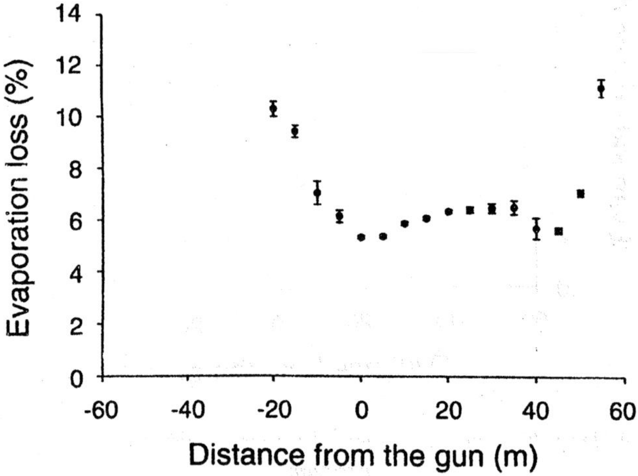

Figure 10. Travelling gun with wind direction at 90° to the average travel direction.

The relative distance shown in Figs. 11 to 14 refers to distance measured along the irrigation pipe with reference to the location of thefirst set of collectors. Compared to the data from the travelling guns, there is hardly any difference in the average spray losses as a function of distance along each row, indicating that the spray droplets coming from the different impact sprinklers along the pipe were relatively uniform. The CV shown in Table I represents all of the data for the centre pivot system. The negative losses shown for two of the tests were recorded under weather conditions that were conducive to condensation rather than evaporation. Whenever the temperature of the source water was lower than the dew point temperature of the surrounding atmosphere, water from the air could have condensed on the spray droplets and caused a gain rather than a loss. The groundwater temperature in this area ranged between 7 and 8EC.

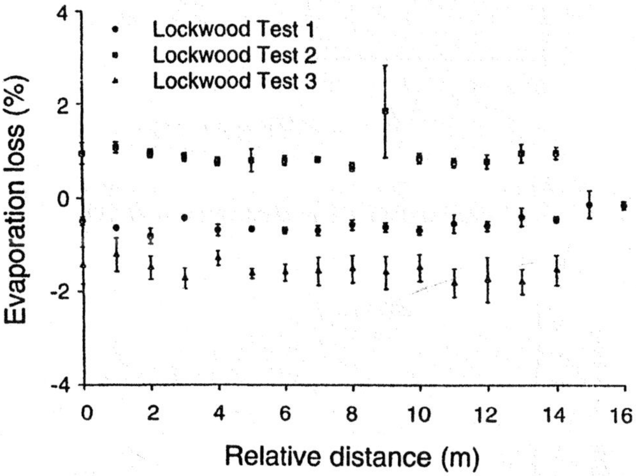

Figure 11. Lockwood 2265 centre pivot with impact sprinklers Nozzle location: 4 m above ground at 2.7 m spacing.

This was lower than the dew point temperatures recorded during some late evening tests. Table I indicates that about four to six collectors/row are sufficient to measure the losses with a precision of ± 0.5%.

Figure 12 shows the results obtained using a Reinke Electrogator 60KT centre pivot system equipped with

Senninger 6Ex 6E 5006-2 two-way impact sprinklers located 4.1 m above the ground at a spacing of 2.9 m. The gains are fairly uniform across the different collectors.

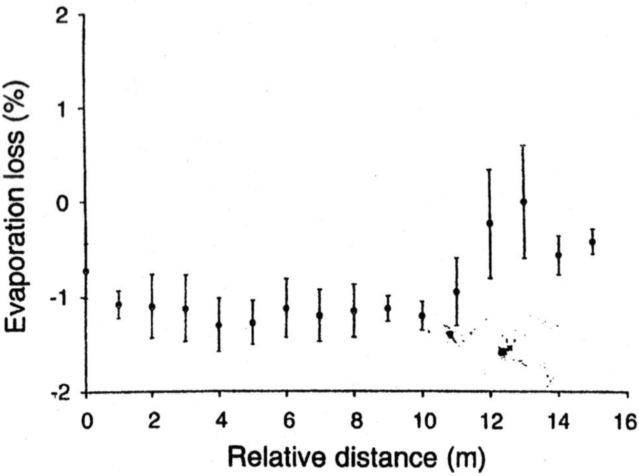

Figure 12. Reinke Electrogator 60 KT centre pivot with twoway impact sprinklers Nozzle location 4.1 m above ground at 2.9 m spacing.

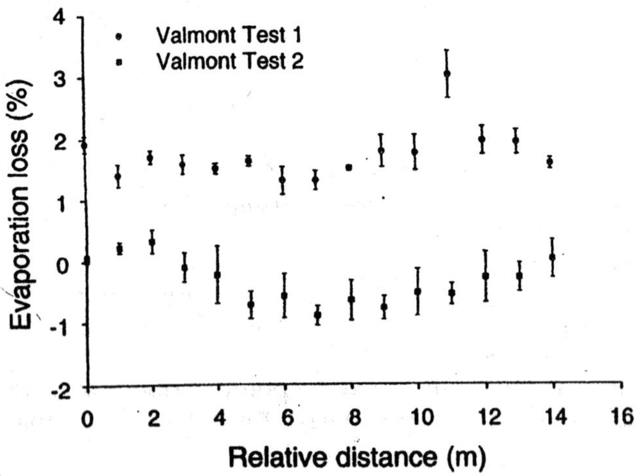

Figure 13. Valmont 4271 centre pivot with flood jets (Nozzle: S.S. Co 3/8 K35) Nozzle location: 3.5 m above ground at 2.9 m spacing.

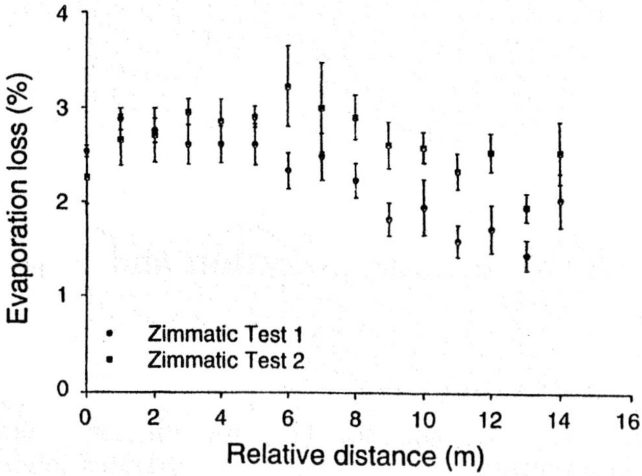

Figure 14. Zimmatic Type 7 Model 510 centre pivot with flood jets (Nozzle: S.S. Co 1/2 K40) Nozzle location: 4.3 m above ground at 2.5 m spacing.

The collector rows were positioned towards the middle portion of the centre pivot pipe.

References

Dadiao, C. and W.W. Wallender, "Droplet size distribution and water application with low-pressure sprinklers", *Transactions of the ASAE*, 1985.

Frost, K.R. and H.C. Schwalen, "Evapotranspiration during sprinkler irrigation", *Transactions of the ASAE*, 1960.

__________, "Sprinkler evaporation losses", *Agricultural Engineering*, 1955.

Myers, J.M., C.D. Baird and R.E. Choate, *Evaporation losses in sprinkler irrigation*, Florida Water Resources Research Center Publication No. 12. University of Florida, Gainesville, 1970.

Pair, C.H., W.H. Hinz, K.R. Frost, R.E. Sneed, and T.J. Schiltz, *Irrigation*, The Irrigation Association, Silver Spring, MD, 1983.

Seginer, I., "Water losses during sprinkling", *Transactions of the ASAE*, 1971.

6

Lawn Sprinkler Selection and Layout

For efficient water application, lawn irrigation systems must apply water uniformly over the entire irrigated surface. High uniformity means that all parts of the irrigated lawn receive about the same amount of water. This requires proper sprinkler selection and layout. Sprinkler patterns of coverage must overlap to provide uniform water application, and pressure must be adequate to provide proper sprinkler operation and water distribution. The layout must direct water into the irrigated zone, but prevent overspray beyond the irrigated zone onto surrounding streets, sidewalks, buildings, or other property. Lawn irrigation systems can often be subdivided into approximately rectangular subunits or zones.

Sprinkler Water Distribution

The pattern of water distribution from a sprinkler is commonly a circle or part circle. Most manufacturers make sprinklers that are easily adjusted to apply water to a complete circle or to almost any portion of a circle. Some sprinklers are pre-set in commonly required patterns such as 90° (quarter-turn) for corners and 180° (half-turn) for the sides of irrigated zones. Figure 1 shows typical The rate of water application from a sprinkler is typically greater near the sprinkler and lower farther from the sprinkler. A very common water distribution pattern is triangular as shown in Figure 2a. Actual sprinkler water distribution patterns will vary from the ideal triangular pattern shown, however, the triangle is an adequate approximation for most cases. The actual patterns can be obtained from the sprinkler manufacturer.

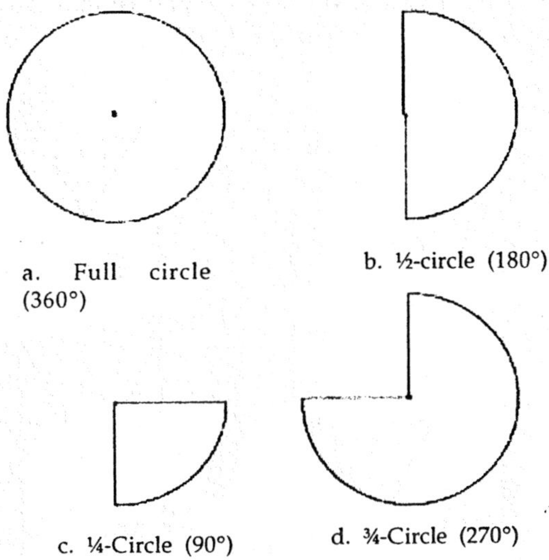

Figure 1.

The sprinkler in Figure 2a has an application rate of 5 mm per hour near the center of the pattern, but that rate drops to zero at a distance of 7.5 m from the sprinkler. This figure shows why sprinkler patterns of coverage must be overlapped in order to obtain uniform applications. If this single sprinkler was allowed to operate for 1 hour, 5 mm of water would be applied near the center of the wetted circle, but that amount would decrease as the distance from the sprinkler increases. As a result, some of the lawn near the sprinkler might be adequately irrigated, but the lawn far from the sprinkler would almost certainly not receive enough water, and uniformity would be poor.

To uniformly apply water with a sprinkler irrigation system, the sprinklers must be spaced so that the water overlaps from one sprinkler to the next (Figure 2b). If the sprinklers are spaced so that the overlap is approximately equal to the radius of coverage (7.5 m in this example), then the resulting cumulative water application rate and depth will be uniform. The dashed line in Figure 2b shows the effect of the overlap from three sprinklers which have individual water application patterns as shown in Figure 2a.

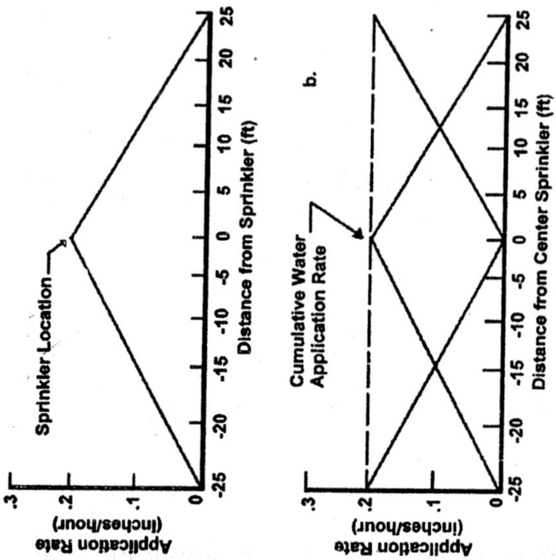

Figure 2

In this case, the cumulative water application is very uniform because the water distribution patterns were assumed to be perfect triangles. In actual systems, the cumulative water application will likely not be this uniform because the water distribution patterns are not perfect triangles.

Sprinkler Layout

To apply water uniformly to square or rectangular land areas with circular sprinkler distribution patterns, sprinklers must be laid out as shown in Figure 3a . In this figure, the land area is a square, and the sprinkler selected has a radius of coverage that is 1/3 of the length of a side of the square. This selection requires 16 sprinklers to uniformly irrigate this area. To uniformly irrigate the corners of this lawn requires four ¼-circle sprinklers. To uniformly irrigate the boundaries requires eight ½-circle sprinklers, two along each side. To provide proper overlap in the center of the lawn requires four full-circle sprinklers.

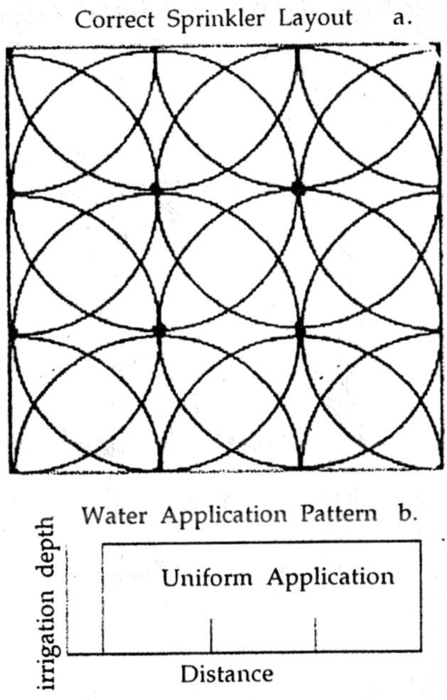

Figure 3

Figure 3b shows the expected water application pattern across the irrigated zone. If sprinklers and flow rates are properly selected, the water application will be very uniform within the irrigated area as shown. If all sprinkler angles are properly adjusted, then water will only be applied up to the edge of the lawn, avoiding overspray onto surrounding streets and sidewalks.

To obtain high uniformity of water application onto square or rectangular land areas with circular sprinklers requires the use of part-circle sprinklers along the edge of the lawn. This rapidly increases the number of sprinklers required because each of the border sprinklers irrigates a smaller land area than the center sprinklers. To reduce the system cost, some systems are installed with fewer sprinklers. However, the initial cost savings from doing this is poor economy because the nonuniform water application will result in inadequate irrigation, waste of water, or both.

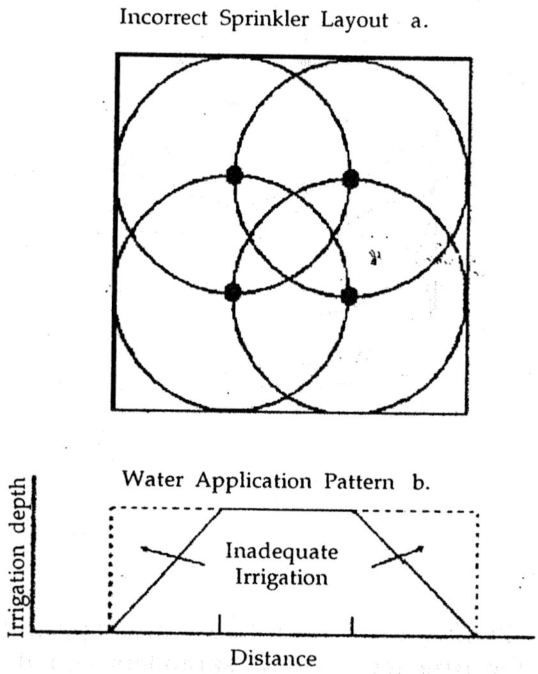

Figure 4

Figure 4 illustrates an attempt to greatly reduce the number of sprinklers required to irrigate, without overspray, the lawn shown in Figure 3 by eliminating all part-circle sprinklers.

It can easily be seen that it is impossible to uniformly irrigate this area without using part-circle sprinklers on the border. In fact, the corners of the lawn will receive no irrigation at all. Figure 5 illustrates an attempt to reduce the number of sprinklers required to irrigate the lawn shown in Figure 3 by using only full-circle sprinklers but allowing overspray onto surrounding areas. The resulting water application pattern will be uniform only between the sprinklers. Irrigation will be inadequate near the lawn boundaries, and water will be wasted from overspray. With this design, the uniformity can be increased by moving the sprinklers nearer the boundaries, but this will also increase the amount of water wasted by overspray.

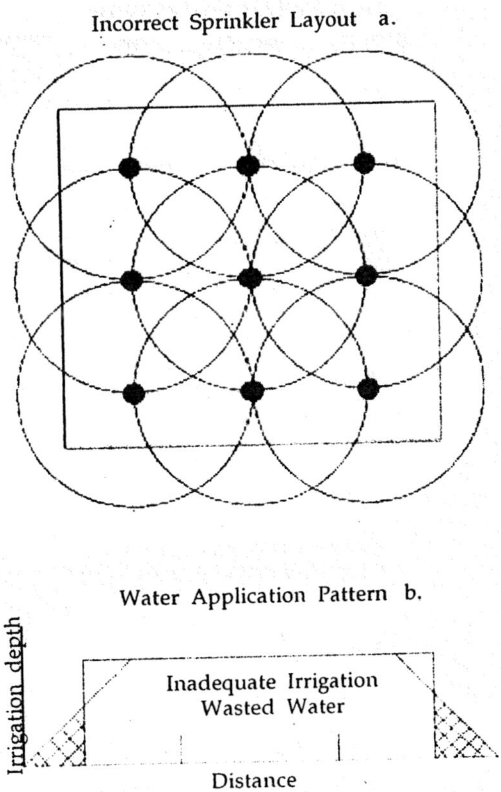

Figure 5

Often homeowners must irrigate long narrow strips of lawn. The correct sprinkler layout for long, narrow lawn areas is shown in Figure 6a. Part-circle sprinklers should be used to border the area and spray toward the center. As shown in Figure 6b, poor uniformity and inadequate irrigation will result from placing the sprinklers in a single line down the center of the strip if overspray is to be avoided. Poor uniformity and the waste of water will result from allowing overspray if sprinklers are placed in a single line down the center of the strip of lawn (Figure 6c).

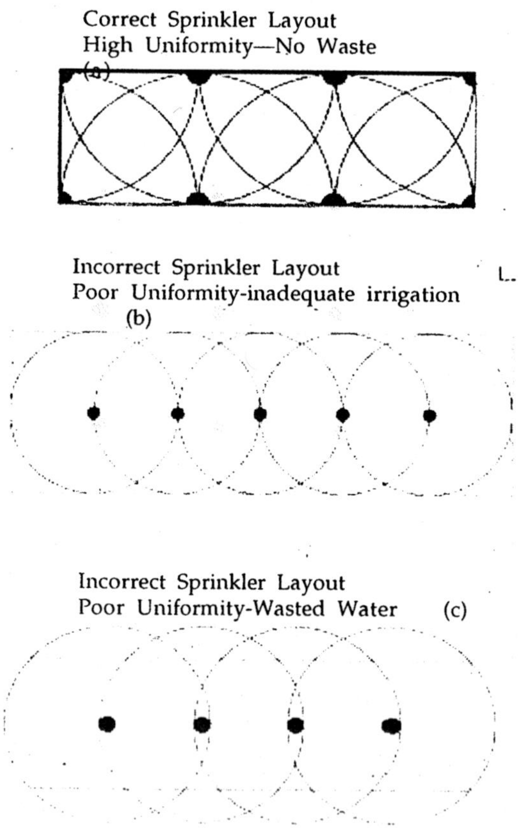

Figure 6

When part-circle sprinklers are placed along edge of a lawn next to a street or driveway, the sprinklers can easily be damaged by vehicle traffic. To avoid this problem, the sprinklers can be moved away from the edge of the driveway a short distance (2-3 ft) into the lawn. Then the sprinkler angle of coverage can be set slightly larger to cover the area near the driveway, without significantly reducing uniformity of water application. Also, always use flexible connectors on sprinkler risers so that pipes will not be broken by occasional vehicle traffic. Flexible connectors will also make it easier to keep sprinklers upright and to make repairs that might occasionally be necessary.

Sprinkler Selection

To achieve uniform water applications from a combination of part and full-circle sprinklers as shown in Figure 3a requires that the sprinkler flow rates be appropriate for the pattern of water distribution. Since a ½-circle sprinkler irrigates only ½ of the land area as compared to a full-circle sprinkler with the same radius, then the flow rate of the ½-circle sprinkler must be only ½ of the full-circle sprinkler flow rate. Likewise, the flow rates of each of the other part-circle sprinklers must be matched to their areas of coverage. For example, a ¼-circle sprinkler must have ¼ of the flow rate of a full-circle sprinkler, etc.

The application rates of part-circle sprinklers are matched to the full-circle rate by changing the sprinkler nozzle size. Manufacturers provide charts which give sprinkler flow rates with different nozzle sizes and operating pressures. Most also provide lists of matched sprinklers for common angles of coverage. In general, however, if the flow rate is to be reduced by ½, then the nozzle cross-sectional area must be reduced by ½. Since the cross-sectional area depends on the square of the diameter, reducing the nozzle diameter by ½ will reduce the flow rate to approximately ¼ of the original flow rate.

References

Isrealson and Hansen, *Irrigation Principles and Practices,* 1967.

Kay, M., *Sprinkler Irrigation: Systems and Practice.* Bedford, 1986.

7

Stationary Sprinkler Irrigation Systems

Sprinkler spacing and design guidelines have been developed primarily for freshwater irrigation with the primary goal of ensuring that those areas of the field receiving the least amount of water receive an adequate amount to sustain the crop and achieve yield goals. To achieve minimum desired application depths within the "lighter application zones," sprinkler spacings of 50 to 65 % of the wetted sprinkler diameter have been determined to be "optimum" to compensate for the declining application along the perimeter. Narrower spacings are typically justified for smaller sprinklers and higher value crops. However, narrower spacings may also result in some zones receiving more water than necessary, and certainly more than the average. A good irrigation design considers these factors and uses a sprinkler spacing that achieves a balance between the relative proportion of "under" and "over" irrigated areas in order to achieve the most uniform application possible.

The application uniformity can be quantified using one of several uniformity indices. The uniformity index recommended for wastewater application is the Christiansen Uniformity Coefficient, U_c. An application uniformity index of 50 is the minimum acceptable for wastewater application using stationary sprinklers. Irrigation systems should be field calibrated regularly to ensure that application uniformity is within the acceptable range. Field calibration can also be used to determine the area within a field receiving an acceptable uniform application.

IRRIGATED ACREAGE

In an effort to answer technical specialists' questions and to provide uniform interpretations of the state's animal waste management rules, the North Carolina General Assembly formed an interagency committee in 1996. The SB 1217 Interagency Committee is composed of two representatives of each of the five agencies with responsibilities for the development and/or enforcement of animal waste management rules. The committee recently adopted guidelines and procedures for determining the irrigated acreage that can count toward the acreage needed to satisfy the land application requirements in the Certified Animal Waste Management Plan (CAWMP).

The committee considered many factors including recommendations from irrigation engineers, certified irrigation designers, and industry representatives before arriving at these guidelines. The irrigated acreage determined by these procedures is intended to "reasonably and practically" account for physical limitations of the application equipment. The "irrigated acreage" computed by the procedures presented below must equal or exceed the acreage requirement specified in the CAWMP for proper nutrient use.

The irrigated area determination includes two broad categories: existing irrigation systems–those systems installed before the guidelines were finalised–and new or expanded irrigation systems installed after the SB 1217 committee released the third revision of the Sixth Guidance Document. Future updates and revisions may occur, so you should refer to the most recent Guidance Document for the latest interpretation and effective dates.

For the purpose of computing the irrigated acreage available to satisfy the CAWMP, the SB 1217 Interagency Committee adopted the term "CAWMP wettable acre" to be applied to existing systems. The irrigated acreage for new or expanded systems should be based on standard irrigation design guidelines, which are based on the effective design area. The term expanded irrigation system applies to new irrigation components that wet an area of a field that was not wetted before adoption of the new guidelines. These terms are defined below.

Existing irrigation systems–For stationary sprinkler systems designed and installed in accordance with standard overlap recommendations (sprinkler spacing between 50 and 70 percent of verified wetted diameter) and laid out with multiple overlapping laterals, the irrigated area allowance is the entire "net wetted area" in the field. The net wetted area is the part of the field that gets "wetted" by one or more sprinklers when operated during normal conditions, i.e., wind speed under 5 mph. The "wetted area" for a single sprinkler operated without overlap is the area inscribed within its wetted diameter as shown in Figure 1.

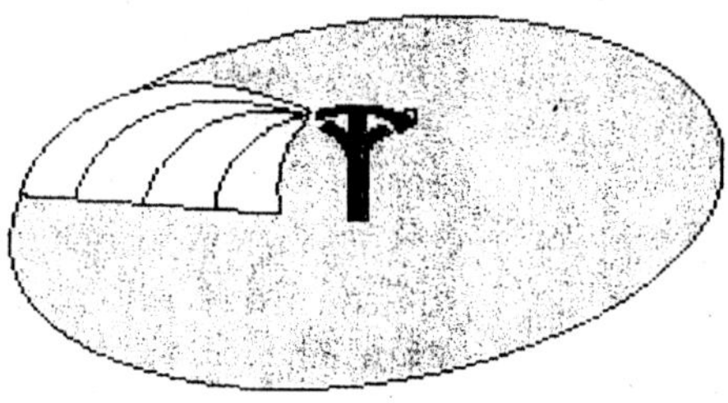

Figure 1. Wetted area of a stationary sprinkler operated without overlap.

For multiple sprinklers such as shown in Figure 2, the entire shaded area gets wetted; however, in this case, the wetted area is referred to as "net wetted area." Due to overlap, some areas are wetted by multiple sprinklers. Obviously these overlap areas cannot be counted twice, hence the term "net" is used. The sprinkler spacing is represented by the inscribed rectangle. For a stationary sprinkler system, there are two sprinkler designations within the field that affect determination of irrigated acreage.

— Interior sprinklers

— Exterior sprinklers

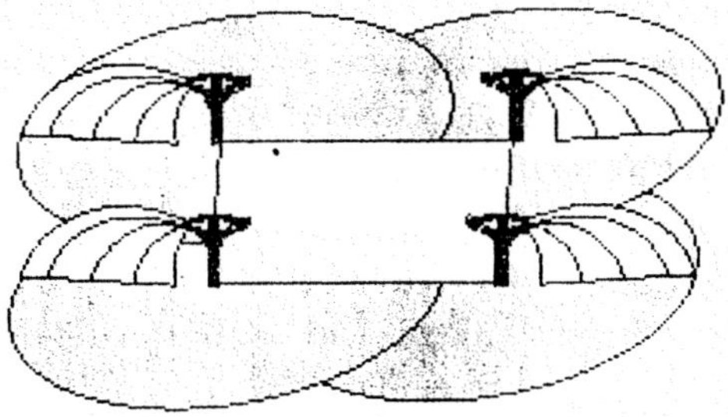

Figure 2. Net wetted area for multiple sprinklers in a square sprinkler pattern.

An interior sprinkler is any sprinkler that receives overlap on all sides. For a rectangular spacing within the recommended spacing range (less than 70 percent of wetted diameter), an interior sprinkler receives overlap from eight adjacent sprinklers as shown in Figure 3, although only four sprinklers contribute significant overlap. An interior sprinkler is a sprinkler that receives overlap from the four adjacent sprinklers along perpendicular transects drawn through the sprinkler in question.

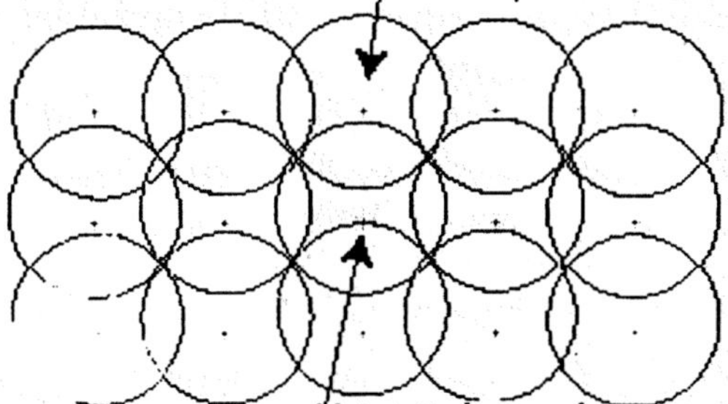

R.lative position of interior and exterior sprinklers.

For the sprinkler pattern shown in Figure 3, only the three center sprinklers are interior sprinklers, and the net wetted area for one interior sprinkler is represented by the center shaded rectangle. The other 12 sprinklers are exterior, and the net wetted area of one exterior sprinkler is represented by the top shaded area. Note that the two shaded areas are not the same size.

For stationary sprinkler systems arranged in a single lateral pattern, the net wetted area should be computed based on 90 percent of the wetted diameter as shown in Figure 4.

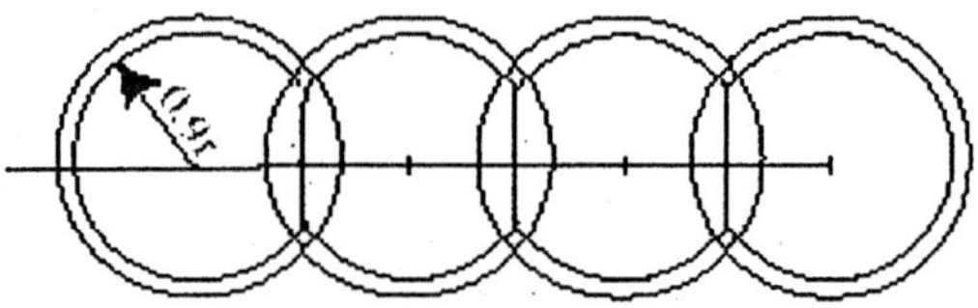

Figure 4. Net wetted area of stationary sprinklers arranged along a single lateral.

The outer portion that does not overlap with an adjacent sprinkler is not included for reasons explained in the next section. For any system in which the lateral spacing exceeds 70 percent of the wetted diameter, each lateral should be treated as a "single lateral" case. If sprinkler spacing along the lateral also exceeds 70 percent of wetted diameter, each sprinkler should be treated as a single sprinkler case. The system layout, including determination of lateral and sprinkler spacing, lateral configuration, and number of interior and exterior sprinklers, must be determined in order to compute the CAWMP wettable acres.

New or Expanded Irrigation Systems

New or expanded irrigation systems should follow recommended design standards, which base the allowable irrigated area on the effectively irrigated area, referred to as the "design area."

The effective irrigated area is the wetted area that receives at least 50 percent of the target application amount. Recent field calibration measurements have determined this to be the area that falls within 78 percent of the wetted radius. Note that application depths remain within 90 percent of the target amount out to 60 percent of the wetted radius. Between 60 and 70 percent of the wetted radius, application amounts still remain within 80 percent of the target application amount. But beyond 70 percent of the wetted radius, application amounts drop off quickly, declining to 50 percent by 78 percent of the wetted radius. Beyond 90 percent of the wetted radius, the application depth drops below 20 percent.

Traditional design spacing guidelines established by the North Carolina Cooperative Extension Service for stationary sprinklers have been 50 to 65 percent of manufacturers' published wetted diameter. When sprinklers are arranged in a square pattern, the point of intersection of the 50 percent application amount occurs at a sprinkler spacing of 68 percent of the wetted diameter.

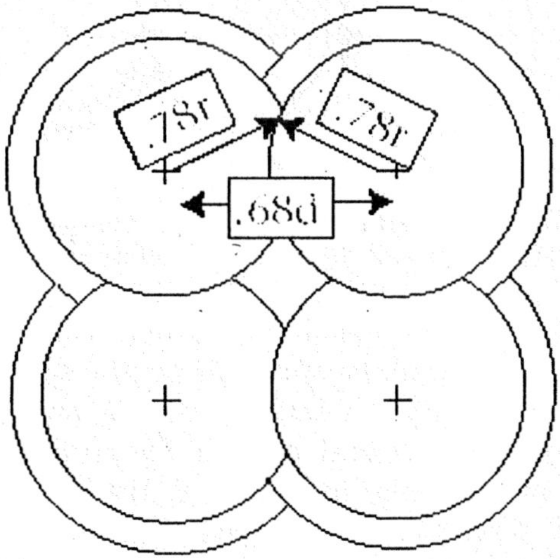

Figure 5. Sprinkler spacing as a percent of wetted diameter where the 50 percent application from adjacent sprinkler overlaps.

This is shown graphically in Figure 5. At wider spacings, there is inadequate overlap resulting in "dry zones." The relative proportion of dry zone (unshaded area in the center of Figure 5) increases as the spacing increases. A conservative rule of thumb for sprinkler spacing has been not to exceed 65 percent of the manufacturer's published wetted diameter. Recent measurements on more than 50 systems determined that field-measured wetted diameters averaged 10 percent less than values published by manufacturers.

A spacing based on 70 percent of field-measured diameter is roughly the same as a spacing based on 65 percent of manufacturers' values. Spacings greater than 70 percent of the verified wetted diameter or 65 percent of the wetted diameter published in manufacturers' literature are considered excessive and result in unacceptable uniformity.

The irrigated area allowance of a single or excessively spaced sprinkler is the area inscribed within 78 percent of the wetted radius as shown in Figure 6.

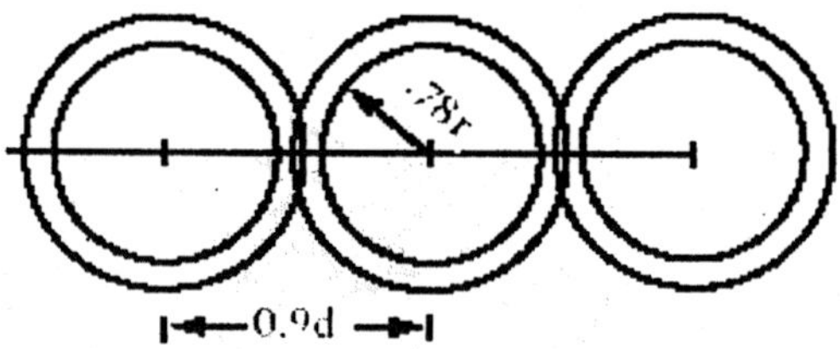

Figure 6. Effective irrigated area of a single sprinkler when sprinkler spacing exceeds 70 percent of wetted diameter.

This is regardless of when the system was installed. You should treat a stand-alone sprinkler or any sprinklers located on laterals in which the lateral spacing and the sprinkler spacing along the lateral exceed 70 percent as a single sprinkler. For example, in Figure 6, the spacing along the lateral is 90 percent of the wetted diameter, so the area of each sprinkler is computed as a single sprinkler rather than as a single lateral as was shown in Figure 4. In Figure 4, the shaded areas overlap, therefore computations are based on a "single lateral." In Figure 6, the shaded areas do not overlap, therefore the computations are based on each sprinkler individually.

Field Measurements

To accurately calculate the irrigated area, determine the wetted diameter or radius of the sprinkler system. There are two methods for determining the wetted diameter, and both require operating the system:

- Directly measure the wetted diameter (footprint), which is the preferred method
- Estimate wetted diameter from field-measured sprinkler pressure and values published in manufacturers' charts for the observed pressure.

Footprint measurement—Footprint measurement involves observing, marking, and measuring the farthest distances from the sprinkler that get wetted. Field data should be collected on at least two sprinklers located on two different lateral lines farthest from the pump. In both cases, the sprinklers should be located at least ¼ of the way down the lateral line. Measurements should be made during very light wind (less than 5 mph). The wetted distance from each sprinkler should be determined at four points along the perimeter as indicated in Figure 7.

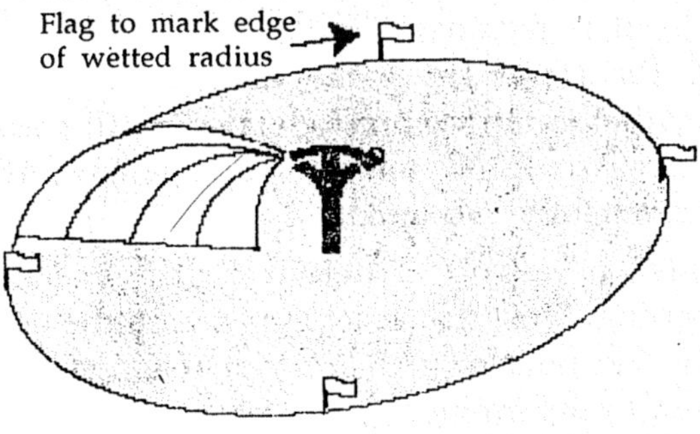

Figure 7. Field determination of sprinkler wetted radius.

The system should be operated long enough for all air to be purged from the system before starting to make measurements. With the system operating at normal pressure:

Pressure measurement—The wetted diameter can also be estimated from pressure measurements if the pressure is measured at the sprinkler. Pump pressure is NOT an acceptable substitute.

Collect field data for at least two sprinklers located on two different lateral lines farthest from the pump. In both cases, the sprinklers should be located at least ¼ of the way down the lateral line. The system should be operated long enough for all air to be purged from the system before measuring the pressure.

If you are using portable quick-connect risers, it is suggested that you configure one riser with a pressure gauge so that you can easily move the entire assembly to several laterals. This riser-pressure gauge assembly should be an "extra" riser and used only when pressure measurements are needed. If it is used routinely, the gauge will soon foul and give erroneous measurements.

For permanent risers, you should install a tee with a threaded reducing port in two riser pipes. You should temporarily install a pressure gauge in the threaded port to make pressure measurements. Once the measurement has been made, it is recommended that you remove the gauge and plug the port. The same gauge can be used at all locations. An alternative approach is to install a shut-off valve between the riser pipe and pressure gauge and leave the gauge permanently mounted.

It is also necessary to determine the exact size of the nozzle opening. Most manufacturers stamp the nozzle size on the end of the nozzle. If this is not readable, a drill index can be used to determine the size of small nozzles (less than ½-inch diameter). Simply insert the shank end of a drill bit into the nozzle opening until a bit providing a snug fit is found. Read and record the size of the drill bit. Once you have measured the operating pressure and nozzle opening, you can estimate the wetted diameter from manufacturers' literature.

REFERENCES

Collier, R.L. and Rochester, E.W., *Water Application Uniformity of Hose Towed Traveller Irrigators*, ASAE Transactions, 1980.

FAO, "Mechanized Sprinkler Irrigation", *FAO Irrigation and Drainage Paper*, 1982.

______________, "Crop Evapotranspiration", *FAO Irrigation and Drainage Paper*, 1998.

8

Center Pivot Sprinkler Systems

A good supply of groundwater and the commercial development of center pivot irrigation systems significantly increased sprinkler-irrigated acreage in many countries, during the late 1960s and early 1970s. Today, center pivot systems, with their automation, large areal coverage, reliability, high application uniformity, and ability to operate on relatively rough topography, are replacing surface, handline, and wheelline systems.

The irrigated area under a center pivot system expands substantially with increasing system length. To accommodate the increased area, the application rate increases linearly along the center pivot lateral through one of two methods: increased flow rates through equally spaced sprinklers or gradually decreased spacing of equal-flow sprinklers along the center pivot lateral. The most common approach is to have equally spaced sprinklers with increasing flow rates (nozzle sizes) along the center pivot lateral.

High-Pressure Sprinklers

In the 1960s, center pivot irrigation systems had standard high-pressure (greater than 50 pounds per square inch) impact sprinklers. These sprinkler packages provided good application uniformity when the system nozzles were properly sized and pressure variation along the lateral was within recommended limits. However, losses from wind drift and evaporation under the dry, windy conditions often encountered in arid and semi-arid environments were excessive.

The sprinkler irrigation industry addressed this problem by developing low angle and low pressure (25 to 40 pounds per square inch) impact sprinklers. These effectively reduced wind drift and evaporation losses, but flow rate variation caused by undulating topography continued to be a significant problem. In the mid 1970s, flow control sprinkler nozzles and fixed pressure regulators were developed. They reduce the flow rate variation due to topography to within tolerable limits. As a result, reduced-pressure impact sprinklers could be used on center pivots.

Low-Pressure Spray Sprinklers

In the mid 1970s, escalating energy costs made the high energy requirement of impact sprinklers a major concern among producers. The sprinkler irrigation industry responded by developing low-pressure spray sprinklers (less than 30 pounds per square inch) for center pivots. These have a fixed-head and a part or fullcircle application pattern. A deflection plate creates spray by deflecting the water jet exiting the nozzle. The deflection plate can be smooth or grooved with a concave, convex, or flat shape. Water leaves the smooth plates as a mist-like spray and leaves grooved plates as tiny streamlets. The sprinklers are either mounted upright on the top of the lateral or mounted upsidedown on drop tubes or booms that extend below the lateral. On undulating topography, pressure regulators are required to minimise flow rate variations and are commonly used to minimise the influence of pressure loss along the lateral.

Spray sprinklers have a smaller wetted area than impact sprinklers and require closer sprinkler spacing. The smaller wetted area greatly increases application rates along the center pivot system. This can intensify runoff problems, particularly on loam and silt-loam soils. Various types of sprinkler booms have been developed to reduce application rates by increasing the wetted area under the center pivot lateral.

Today, the most popular type is an offset boom with a horizontal length of 10 to 20 feet perpendicular to the center pivot lateral. These offset booms are commonly used on the outer one-half to one-third of a center pivot lateral. Recently developed moving-plate spray sprinklers also decrease application rates by increasing wetted area. These sprinklers, such as Rotators, Spinners, and Wobblers, reduce the number of water streamlets which inçreasing drop size and water throw distance. At the same time, they maintain good application uniformity. Moving-plate spray sprinklers combined with offset booms along the outer spans of the center pivot provide efficient irrigation.

LEPA Systems

In the early 1980s, a low pressure application package for center pivot systems known as LEPA (Low Energy Precision Application) was developed for the southern plains states. A LEPA package has very-low-pressure (6 to 10 pounds per square inch) bubblers or furrow drag socks suspended on drop tubes at a height of 1 to 3 feet above the soil surface. Crop rows are planted to follow the circular path of the center pivot system, and alternate furrows are wetted. LEPA systems have characteristically high application rates that usually exceed the water infiltration rate. Basin tillage is required to provide soil-surface storage until the water infiltrates. Some LEPA applicators can be converted to spray heads having wetted areas on the order of 10 to 25 feet in diameter. These have good sprinkler pattern overlap and apply water uniformly. When used in the crop canopy, the heads are usually spaced to match alternate crop rows.

Irrigation application efficiencies of 90 to 95 percent have been measured using LEPA sprinkler packages. This efficiency is the result of reduced evaporation. By locating the applicators within the crop canopy and near the soil surface, the amount of wetted soil and wetted plant surface area is minimised. Wind drift and spray evaporation are also eliminated. However, their high application rates and their limited clearance of the applicators make the LEPA packages unsuitable for slopes.

Application Rate

The main disadvantage of center pivot irrigation systems is the high water application rates under their outer spans. Since sprinkler flow rate increases linearly along the system lateral, application rates at the outer end also increase with the length of the system. Application rates under the outer spans of the standard quartermile-long low-pressure center pivot normally exceed infiltration rate and result in runoff.

Runoff, the lateral redistribution of applied water, causes areas of excessive and deficient soil water content in the field, reducing crop yield and quality in these regions. The potential for localised chemical leaching from the crop root zone also increases in places where runoff collects. Soil-surface water storage in small, natural depressions decreases the actual volume of runoff. Surface storage can be enhanced by tillage practices, such as basin or reservoir tillage.

Infiltration rate, which determines the potential for runoff, is dynamic. Infiltration rate decreases during irrigation (Figure 1). The initial soil water content also affects the infiltration rate; an increase in the initial soil water content decreases the infiltration rate. In addition, infiltration rates normally decrease over the season due to soil-surface sealing from sprinkler droplet impact.

As a result, in row crops such as potatoes, runoff may increase throughout the season. Decreasing infiltration rates combined with high water application rates make runoff a near certainty for standard quarter-mile-long center pivots on all but sandy soils. Optimal center pivot system performance requires the use of both proper sprinkler packages to minimise water application rates and basin or reservoir tillage to minimise runoff.

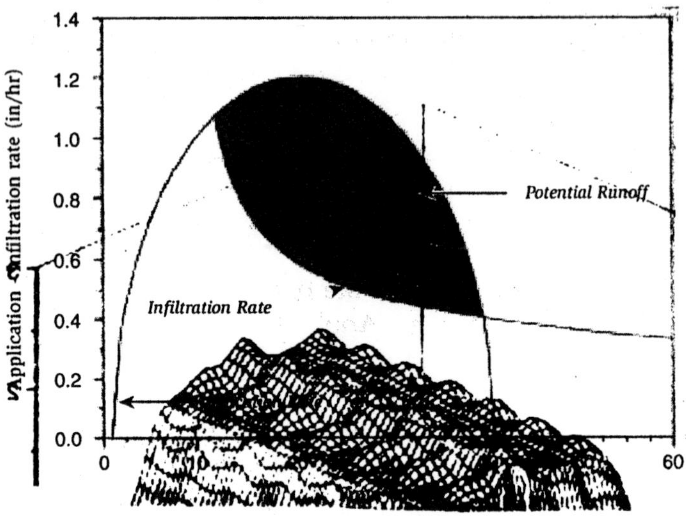

Figure 1. Graphical representation of how water application rates under a center pivot exceed infiltration rate.

Typical relative water application rate patterns for various center pivot sprinkler packages are shown in figure 2. High-pressure impact sprinklers have the lowest application rates followed by lowpressure impact sprinklers. Low-pressure spray sprinkler packages, listed from lowest application rate to highest, are offset booms with rotators, offset booms with sprays, drop tubes with rotators, drop tubes with sprays, and in-canopy sprays.

The peak application rate along the outer spans of a standard quarter-mile-long center pivot system for all the sprinkler packages exceeds the infiltration rate of most soils. Booms are an effective means for increasing sprinkler wetted area while decreasing water application rate. Since application rates are lower nearer the center pivot point, booms are usually only used on the outer one-half to one-third of a quarter-mile-long center pivot system.

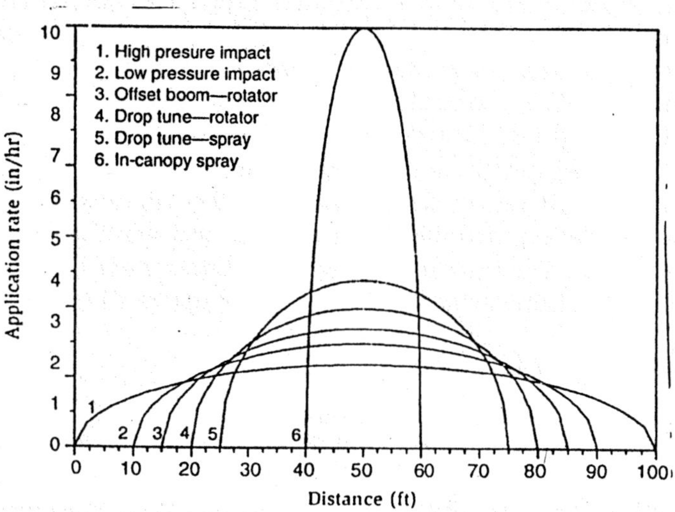

Figure 2. Comparison of relative application rates under various center pivot sprinkler packages.

Low–Pressure Sprinkler Patterns

For a low-pressure center pivot sprinkler package, the shape of the application rate pattern is defined by pressure, nozzle size, plate configuration, sprinkler height, and wind speed. Sprinkler application rate pattern and spacing determine application uniformity.

Pressure and Nozzle Size

Pressure and nozzle size control the drop size distribution from a sprinkler and drop size influences the application rate pattern. Higher pressure creates smaller drops while bigger nozzles produce larger drops. Drop size also influences the trajectory of a given sprinkler droplet. When initial velocities are equal, large droplets will travel farther from the sprinkler than small droplets. Consequently, high pressure or small nozzle sizes, which tend to produce smaller droplets, increase application rates near the sprinkler while low pressure or large nozzle sizes, which tend to produce larger droplets, increase application rates farther from the sprinkler. Obtaining suitable application rate patterns is dependent on following the manufacturer's nozzle size and pressure range recommendations. However, donut application rate patterns may be accentuated at the lowest recommended pressure, reducing application uniformity. At the highest pressure recommendation, droplet size is smaller and wind drift losses will increase. The best results are often found near the middle of the manufacturer's recommended pressure range.

Deflection Plate Configuration

Sprinkler deflection plate configuration has a large effect on the sprinkler application rate pattern. In general, smooth deflection plates produce small drop sizes, which are highly susceptible to wind drift losses, except at lower pressures (10 to 15 pounds per square inch). Serrated deflection plates have many small grooves and are used with fixed-plate sprinklers. Grooved deflection plates have four to six large grooves and are used on moving-plate sprinklers. Moving-plate sprinklers are the most common type.

They maximise wetted sprinkler area while minimising operating pressure. The application rate pattern depends on the number of grooves, trajectory angle, and speed of motion. The number of grooves in the plates affects the drop size distribution. Fewer grooves produce larger streamlets and larger drop sizes, which travel farther from the sprinkler and maximise wetted area. Within limits, greater trajectory angles produce more uniform application rate patterns. The primary disadvantage of higher trajectory angles is a greater susceptibility to wind drift. Lowering the sprinkler elevation will reduce wind drift.

The effect of plate configuration and motion on sprinkler application rate pattern is shown in figures 3 through 7. A 4-groove plate with an 8 degree trajectory produces a concentrated application of water near the outer spans of the wetted pattern, creating a donut-shaped application rate pattern. The application rate pattern for the same sprinkler with a 6-groove plate and a 12 degree trajectory angle creates smaller droplet sizes and increases water application near the sprinkler. The smaller droplet sizes combined with the higher trajectory angle reduce the wetted area slightly. The donut-shaped application rate pattern remains but to a lesser degree because a larger percentage of the water is applied near the sprinkler.

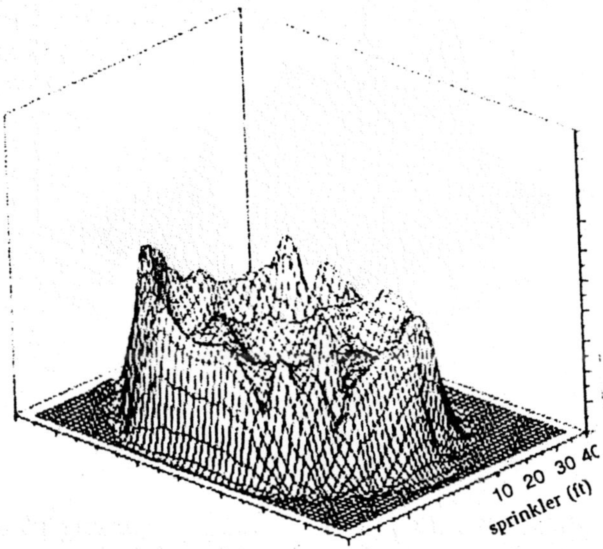

Figure 3. Application rate pattern from a 4-groove rotating-plate spray sprinkler with an 8° trajectory angle.

The application rate pattern for a fast rotating-plate (spinner) with 6 grooves and 12 degree trajectory angle is shown in figure 5. The faster rotation of the plate provides a more uniform application rate pattern of elliptical shape with the highest application rate near the sprinkler. The application rate pattern for the same sprinkler with a 20 degree trajectory angle. The greater trajectory angle slightly increases the wetted area of the sprinkler, reducing the application rate near the sprinkler.

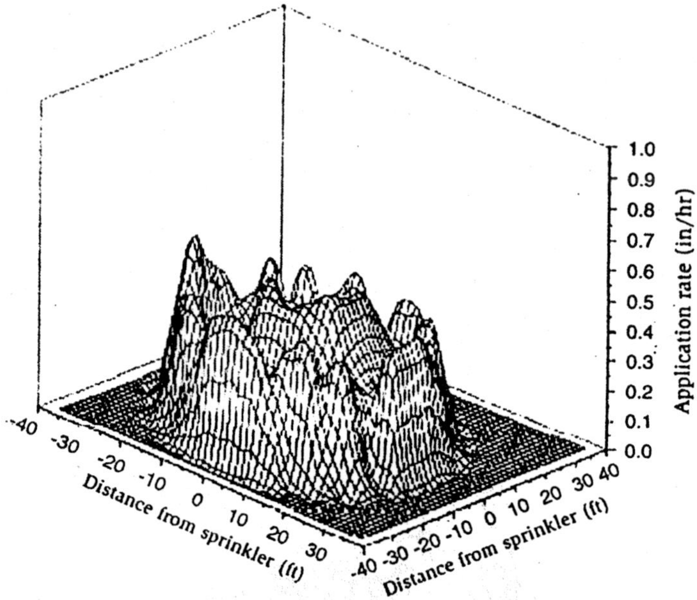

Figure 4. Application rate pattern for a 6-groove rotating-plate spray sprinkler with a 12° trajectory angle.

The application rate pattern from a wobbling-plate type sprinkler having 9 grooves and a 15–degree trajectory angle is shown in figure 6. This application rate pattern resembles a truncated cone with an additional elliptical shaped peak near the sprinkler. The application rate pattern is very uniform except near the sprinkler. For donut-shaped application rate patterns, such as those illustrated in figures 3 and 4, the cumulative application rate pattern produced by multiple sprinkler overlap is reasonably uniform. This, combined with the effect of averaging the cumulative application rate pattern as a center pivot passes over a point on the soil surface, provides excellent application uniformity. Application rate patterns that are more uniform in shape.

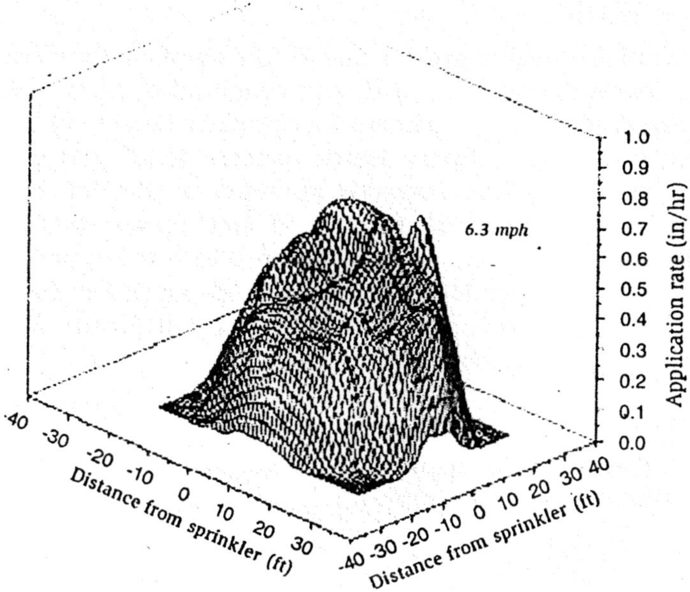

Figure 5. The application rate pattern for a fast rotating-plate (spinner) with 6 grooves and 12 degree trajectory angle

However, the individual sprinkler wetted areas are usually smaller so the required sprinkler spacing is about the same as that of sprinklers with larger donut-shaped application rate patterns.

Sprinkler Height

Sprinkler height influences the size of the sprinkler wetted area and wind drift losses. Increasing sprinkler height increases sprinkler wetted area slightly with no significant effect over the practical heights of 6 to 10 feet. Sprinkler heights greater than 6 feet on short crops (height less than 3 feet) do not significantly increase application uniformity. However, sprinkler heights less than 6 feet significantly decrease application uniformity, particularly for sprinklers having deflection plates with low trajectory angles. With taller crops, the optimal sprinkler height is the maximum canopy height.

Sprinkler heights greater than 6 feet significantly increase spray losses due to wind drift and evaporation. Spray losses average about 3 and 5 percent for sprinkler heights of 3 and 6 feet, respectively. Spray losses increase to 10 percent for sprinklers (spray and impacts) mounted on the top of the center pivot at heights of 12 to 15 feet. Spray losses can double as wind speed increases from 0 to 5 miles per hour to 5 to 10 miles per hour. For short crops, sprinkler heights near 6 feet provide good application uniformity while maintaining reasonable spray losses.

Wind Speed

Wind distorts the application rate pattern from spray sprinklers and affects application uniformity.

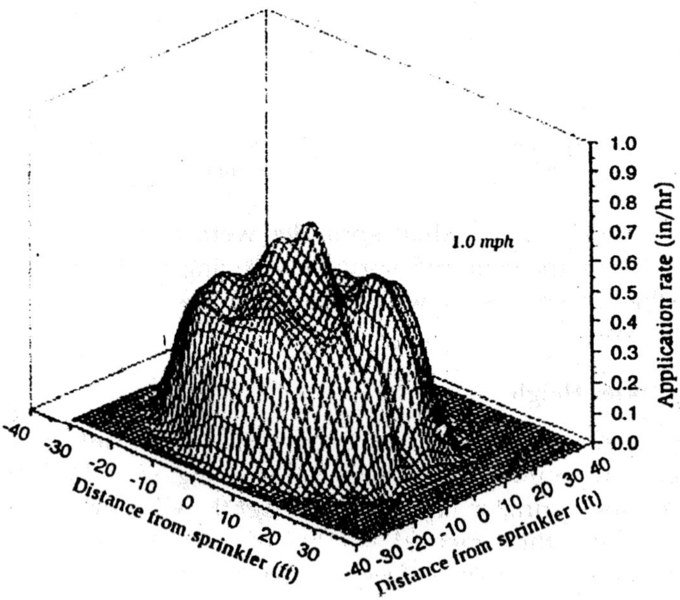

Figure 6. The application rate pattern from a wobbling-plate type sprinkler having 9 grooves and a 15–degree trajectory angle

The effects of wind on the application rate patterns for a Spinner and a Wobbler type spray sprinkler are depicted in figure 7. Comparing these patterns with those of figures 7 for the same sprinklers under lower wind speeds reveals that the application rate patterns are largely shifted downwind.

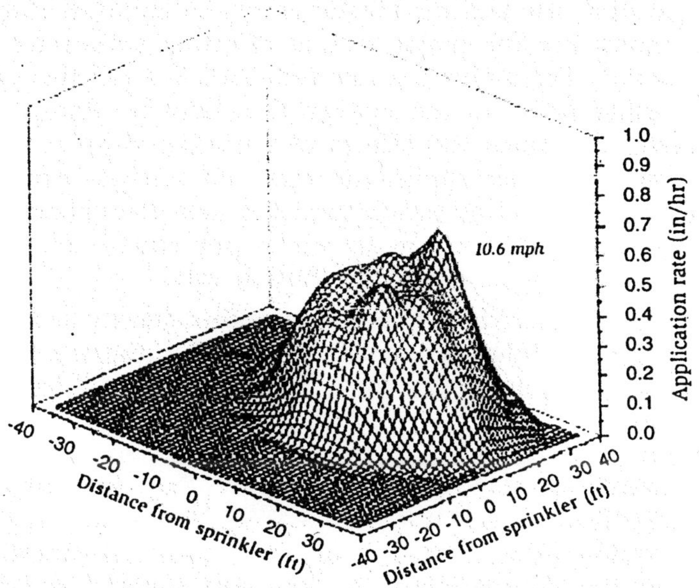

Figure 7. The effects of wind on the application rate patterns for a Spinner and a Wobbler type spray sprinkler are depicted

Distortion of the application rate pattern is most pronounced near the sprinkler where the smallest droplets occur. Computer simulation of composite wind-affected application rate patterns under a center pivot indicates that application uniformity is not significantly reduced for wind speeds up to 10 miles per hour. This favourable result is largely due to the multiple sprinkler overlap required to obtain good uniformity with low-pressure sprinklers and to limiting sprinkler height to about 6 feet.

Sprinkler Droplet Kinetic Energy

Many soils, particularly those containing significant silt fractions, are susceptible to soil-surface sealing from sprinkler droplet impact. The force of the droplets hitting the ground breaks down the surface soil structure, forming a thin compacted layer that greatly reduces infiltration rate. The application rate and the kinetic energy of sprinkler droplets at impact are the major factors affecting soilsurface seal formation. The infiltration rate reduction is a function of the particular soil and the energy flux density. Energy flux density combines the effects of sprinkler droplet kinetic energy and water application rate into a single parameter that is expressed as power per unit area (feet-pounds per minute per square foot or watts per square meter). It correlates very well with infiltration rate.

The relationship between energy flux density and depth of infiltration prior to runoff is illustrated in figure 8 for two different soils under dry, bare conditions. The silt loam soil is very susceptible to soil-surface sealing. The infiltration depth prior to ponding decreases very rapidly with a minimal increase in energy flux density. The loam soil is less susceptible to soil-surface sealing, but the depth of infiltration prior to runoff still decreases significantly as energy flux density increases. The effect of sprinkler droplet impact on the infiltration rate of a particular soil must be measured to develop a quantitative relationship similar to that of figure 8. This is difficult because the results depend on soil surface conditions, soil structure, and soil water content. However, the general trend shown in figure 8 is applicable to any soil and useful in the selection of sprinklers for a center pivot irrigation system.

Studies of runoff under center pivot irrigation systems indicate that soil-surface sealing continues to develop with each additional irrigation. The only way to recover from soil-surface seal formation is to physically destroy it with a tillage operation. The best approach for limiting soil-surface seal formation is to protect the soil surface through residue management and to exclude water application from bare soil conditions.

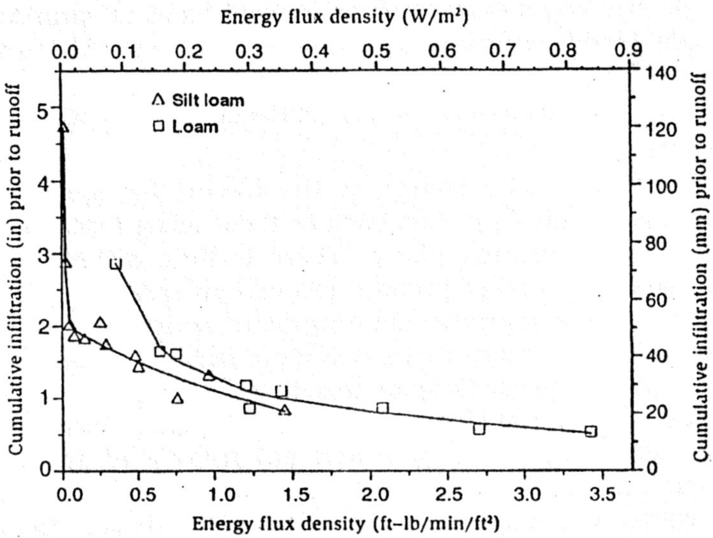

Figure 8. Infiltration rate reduction by energy density flux of sprinkler droplets for two soils

When water applications must be made on bare soils, the energy flux density should be reduced to delay formation of the soilsurface seal. This can be accomplished by either using sprinklers with reduced droplet kinetic energy, reducing application rate, or both. Reducing the application rate is easiest and can be done by renozzling the center pivot system to reduce flow rate. The application rate under a center pivot is independent of system speed, so adjusting the system speed does not affect formation of a soilsurface seal.

The kinetic energy of a sprinkler droplet depends on droplet size (mass) and velocity at impact with the soil surface. Droplet velocity is also a function of drop size. Drop size distribution is determined by sprinkler nozzle size, pressure, and deflection plate configuration.

Figure 9 shows the kinetic energy per unit volume of water applied (foot-pounds per cubic foot or joules per kilogram) versus the dimensionless ratio (ft/ft, m/m) of nozzle size to pressure head for several types of sprinklers. Droplet kinetic energy is highest for sprinklers producing the largest drop sizes, such as standard impact sprinklers and rotator type sprinklers having deflection plates with few grooves.

Droplet kinetic energy is the lowest for sprinklers producing small drop sizes such as those using fixed sprays with flat or serrated plates. There is little difference in droplet kinetic energy between the various spray sprinklers, except for the 4-groove rotating-plate sprinkler. Overall, droplet kinetic energy varies only by a factor of three across all sprinkler types. Despite this limited range in droplet kinetic energy, a study of sugar beet emergence comparing sprinklers with 105 ft-lb/ft^3 and 315 ft-lb/ft^3 of drop- let kinetic energy found a 13 percent increase in sugar beet emergence under the sprinkler with two thirds less droplet kinetic energy. Sprinkler selection does influence soil-surface seal formation. This not only affects infiltration rate, but has other agronomic implications such as soil erosion, water application efficiency, and nutrient distribution in the soil profile.

Optimal Sprinkler Package Selection and Installation

Sprinkler selection and installation have a significant effect on the performance of a center pivot irrigation system. Both application rate relative to infiltration rate and the susceptibility of the soil to surface sealing need to be considered in the system design. The application rate of low-pressure spray sprinklers can be reduced by using offset booms on alternate sides of the center pivot lateral. On soils with extremely low infiltration rates or with a high susceptibility to soil-surface sealing, offset booms on both sides of the center pivot lateral can be used at each sprinkler outlet to further reduce application rate.

The effectiveness of offset booms for reducing application rate is shown in figures 10, 11, and 12. The composite application rate for 6-groove rotating-plate sprinklers on drop tubes is shown in figure 10. Figure 11 shows the composite application rate under the same sprinkler conditions with offset booms on alternate sides of the center pivot lateral. The average application rate is reduced about 30 percent by offset booms.

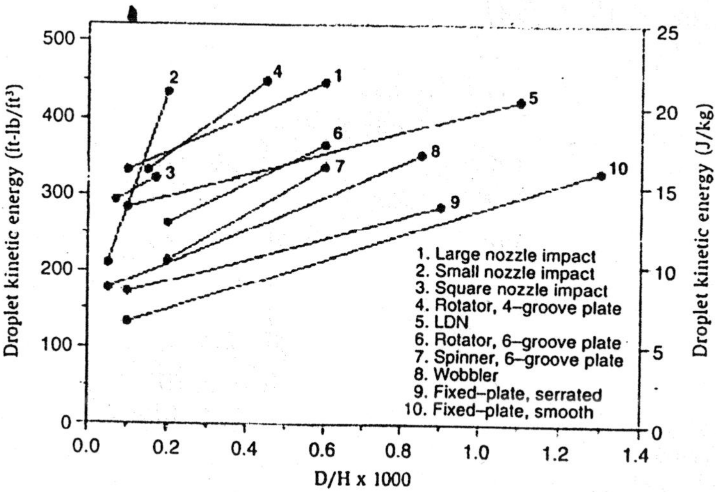

Figure 9. Sprinkler droplet kinetic energy for various sprinkler types as a function of the dimensionless ratio of sprinkler nozzle diameter to sprinkler pressure head.

The composite application rate with two offset booms at each sprinkler location and each sprinkler nozzle providing one-half the flow rate is shown in figure 12. The application rate is reduced 5 percent compared to the single offset boom. The major advantage of the double offset boom is that it uses smaller nozzles, which reduces the kinetic energy of the droplets.

Table 1 lists the average and highest 10 percent application rates for various types of spray sprinklers on offset booms installed on alternate sides of a center pivot lateral.

Table 1. Application rates and application rate reduction provided by offset booms of various lengths with a 10-foot sprinkler spacing and flow rate of 10 gallons per minute.

		Application rate		Application rate		
Sprinkler type	*Offset distance (ft)*	*Average (in/hr)*	*High 10% (in/hr)*	*Average (%)*	*High 10% (%)*	*Application uniformity (%)*
Fixed-plate	0	2.03	3.84	—	—	97
smooth	10	1.45	3.31	71	86	97
	15	1.28	2.56	62	67	97
	20	1.11	2.52	55	66	97
Fixed-plate	0	2.13	4.35	—	—	98
serrated	10	1.52	3.51	71	81	98
	15	1.32	2.87	62	66	98
	20	1.15	2.75	54	63	98
Rotator	0	1.44	2.37	—	—	98
4-groove	10	1.12	2.24	78	95	98
	15	1.00	2.21	70	93	99
	20	0.91	1.82	63	77	98
Rotator	0	1.54	2.47	—	—	97
6-groove	10	1.17	2.27	76	92	97
	15	1.04	2.12	67	86	97
	20	0.94	1.65	61	67	97
Spinner	0	1.58	2.76	—	—	97
6-groove	10	1.22	2.49	78	90	97
	15	1.09	1.97	69	72	97
	20	0.97	1.50	62	54	97
Wobbler	0	1.42	2.41	—	—	100
low angle	10	1.11	2.27	79	94	100
	15	1.00	1.94	70	80	100
	20	0.90	1.41	64	58	100

Offset booms are relatively inexpensive and very effective in reducing the application rate. Since the application rate under low-pressure spray sprinklers can be minimised by using offset booms, sprinkler selection should be based on drop size distribution.

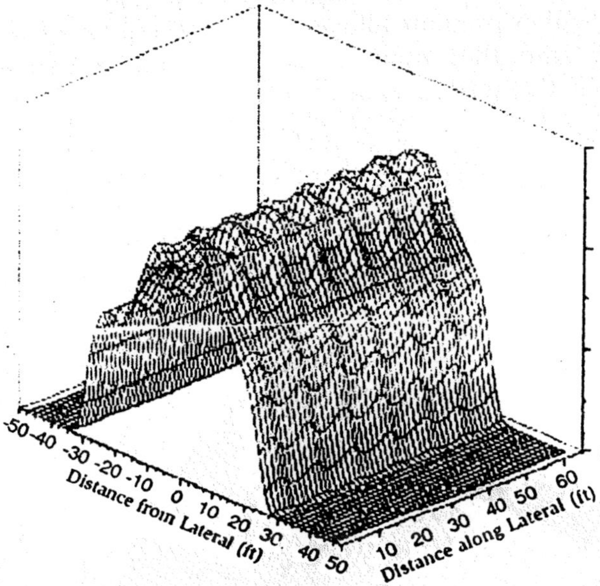

Figure 10. Composite application rate pattern under a center pivot from 6-groove rotating-plate sprinklers on drop tubes with 10-foot sprinkler spacing and 10 gallons-perminute flow rate.

Small drop sizes have the least droplet kinetic energy but are the most susceptible to wind drift losses. Large drop sizes have the highest droplet kinetic energy but are the least susceptible to wind drift losses. Sprinklers that provide a compromise between these two extremes are best. Most movingplate sprinklers have medium drop sizes and maximum wetted area. Because they all have about the same droplet kinetic energy, the final selection of the brand rests on personal preference.

The significant differences in the application rate patterns of the various moving-plate sprinklers influence the spacing of the sprinkler heads. Fixed-plate spray sprinklers with their smaller wetted area require closer spacing than the moving plate spray sprinklers. Wobbler type sprinklers with their more uniform application rate pattern allow for larger spacing. Pressure also has a significant effect on the required spacing. Higher pressure allows wider spacing because of the resulting smoother application rate pattern and slight increase in the wetted area.

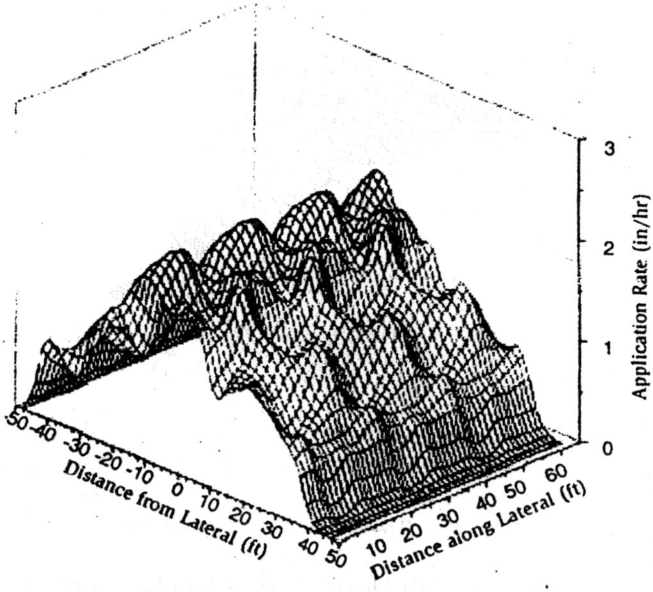

Figure 11. Composite application rate pattern under a center pivot from rotating-plate sprinklers on offset booms having a 15-foot horizontal projection on alternate sides of the center pivot laterai with 10-foot sprinkler spacing and 10 gallons-per-minute flow rate.

With most spray sprinklers, low pressure produces a donutshaped application rate pattern. As a result, closer spacing is needed in order to maintain application uniformity. Due to the high flow rates required on the outer portion of center pivots, large spacings require large nozzle sizes, which may result in excessively large drops, particularly at low pressures.

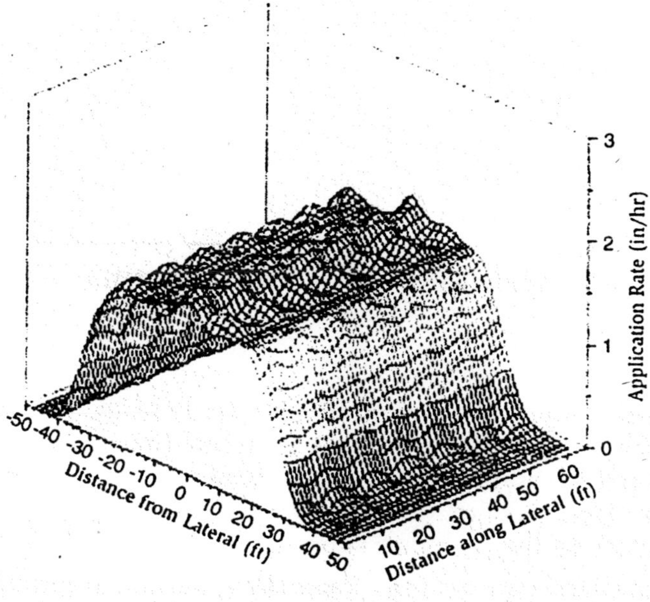

Figure 12. Composite application rate pattern under a center pivot from rotating-type sprinklers on an offset boom having a 15-foot horizontal projection on both sides of the center pivot lateral with 10-foot sprinkler spacing and 10 gallon-per-hour flow rate.

Center pivot sprinkler outlets are normally spaced about 8 to 10 feet apart. This spacing is adequate for all but fixed-plate spray sprinklers and rotators at 10 pounds per square inch. Since every sprinkler outlet is normally used along the outer half of a standard quarter-mile-long center pivot, all the moving-plate type spray sprinklers provide good application uniformity. In general, all moving-plate type sprinklers provide good application uniformity with spacings normally encountered on center pivots.

References

Kincaid, D.C., *Spray drop kinetic energy from irrigation sprinklers*, 1996.

Kincaid, D.C., M. Nabil, and J.R. Busch, "Spray losses and uniformity with low pressure center pivots", *ASAE Paper*, 1986.

9

Wheelmove Sprinkler Systems

The term "wheelmove" seems to be the industry preferred label for what are also known as wheel-line, sideroll, or lateral-roll irrigation machines. A wheelmove consists of the mover, lateral pipe, wheels, sprinklers, couplers, and connectors to the mainline supply.

The lateral pipe conveys water along the line to sprinklers typically spaced 30 to 40 feet apart. While system lengths vary, most are 1280 feet long to match a quarter mile wide field. The lateral pipe also acts as an axle for wheels midway between the sprinklers. The lateral pipe is typically either 4 inch or 5 inch diameter, with some systems using 5 inch close to the mover and 4 inch further away. The 5 inch pipe is desirable as a stronger axle where the system operates on a side hill or on undulating ground.

The wheels vary in diameter from five feet (actual diameter is 64 inches) to ten feet. The most common wheel diameter is 7 foot (76 inches) which moves the line 60 feet laterally with 3 revolutions. Five feet (64 inches) wheels are cften used to give a 50 feet move with three revolutions. Probably a significant majority of wheelmoves have 7 feet diameter wheels with 5 inch pipe.

The height of lateral line/axle limits the crops which are adapted to wheelmove systems to approximately half the wheel diameter. It is not likely to see a wheelmove in a corn field except, perhaps, for one or two early irrigations, nor would it be used in orchards or any other similar tall crops. Nearly 80% is in perennial forages including alfalfa, hay, and pasture.

The couplers connect the lateral pipe sections together and transmit the axle torque. They also include a drain and an outlet for the sprinkler. In fields of varying width, such as those at an angle to a roadway, the ease of dropping or adding lateral sections is important. Thus, a clamp type coupler may be desirable. Also, the use of a leveler on the sprinkler is not only recommended but is an essential requirement for any wheelmove system.

The joint of pipe making up the lateral pipeline serve as the axle for the wheelmove system and are supported above the ground by wheels mounted midway along their length. Sprinklers are located at the connections and ends of the lateral pipe, positioning a sprinkler head midway between each well and at the ends of the lateral pipeline. Normally, a wheelmove system has impact sprinkler heads with levelers to keep the sprinklers in an upright position.

Located adjacent to each sprinkler is a drain that automatically empties the lateral pipeline when water pressure drops off. The couplers between the pipe joints provide a watertight connection and also transmit torque, produced by the power mover, to the entire length of the system. The power mover is mounted in the center of the wheelmove system and provides the power to move or drive the irrigation system. One end of the lateral line is connected via a flexible hose to a pressurised mainline pipe with risers at suitable intervals.

Mechanical Operation

At periodic intervals (typically, 11 ½ or 23 ½ hours) the system is rolled 50 or 60 feet laterally by a small single cylinder gasoline or diesel engine power unit. The power unit, (or power mover) operates through a combination of hydraulics, gears, and chains located at the middle or end of the lateral. All the water must be drained through the automatic drains from the line prior to moving. The move usually takes 20 to 30 minutes, including draining, which is considerably less time than required for a hand move line.

Prior to moving the line it should be straight. Then the mover will be slightly ahead of the line, which will be tight without slack when moving. If the line migrates to one end in moving, the opposite end should be "trailed" back manually. This unbalance will tend to shift the line in the desired direction. The end should be realigned after the line position is corrected. It is important to keep the wheels from getting ahead of the mover so as to prevent serious misalignment.

When the wheel line operates crosswise on a slope, the line may migrate down slope. If the mover is situated off center towards the downhill side, the extra "pull" from the longer uphill line will tend to offset this downward migration.

Irrigation Application

A typical wheelmove uses impact sprinklers with levelers and has 3/16" diameter nozzles operating at 45 to 50 psi pressure. Unfortunately, some systems are operating at pressures somewhat below this which lead to problems with distribution uniformity.

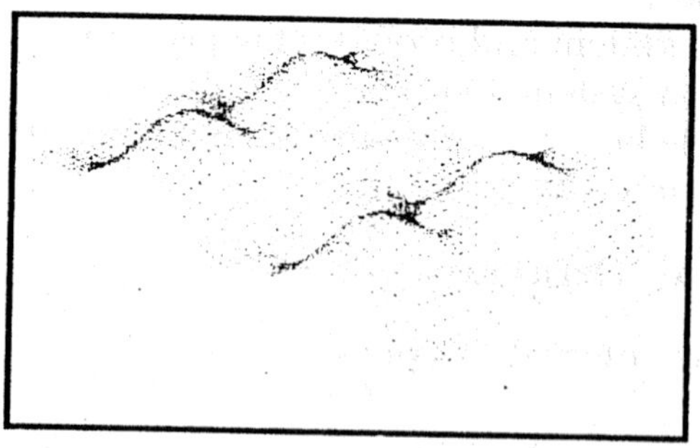

Figure 1. Wheelmove water distribution, no offset, CU of 62%.

Irrigation application efficiencies (water stored in the root zone divided by a water delivered to the field) for wheel moves varied from 39% to 75%. With reasonable good management, the wheelmove application efficiencies averaged nearly 70%. The lower application efficiencies were from lower pressures than designed or, perhaps, a greater distance in the move than is desirable. One such system we evaluated had a coefficient of uniformity (CU) of 62% on the 40 foot by 60 foot sprinkler spacing (Fig. 1).

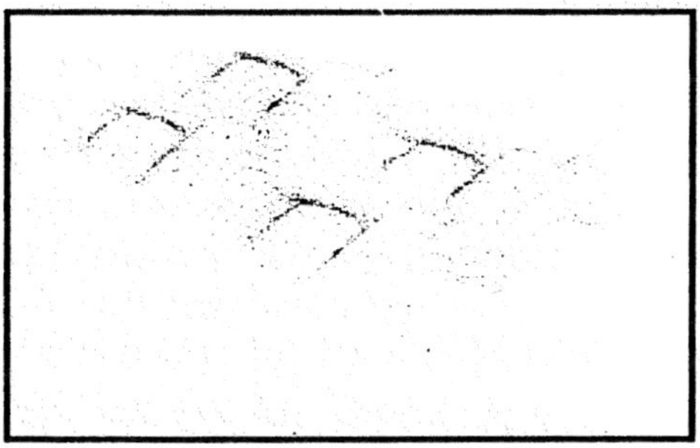

Figure 2. Wheelmove water distribution with 20 foot offset in 2nd irrigation (CU of 87%).

The uniformity improved to 87% when a 20 foot offset (Fig. 2) was used on alternate irrigations. The 20 foot offset, in this case, was equivalent to one revolution of the wheel move to either the right or the left of the previous irrigation lateral position.

Wheelmove Sequencing

The objective of the moving sequence is to maintain adequate soil water for best crop growth, while not causing excess deep percolation. Deep percolation is detrimental to water conservation and salinity reduction goals. Various wheel move sequencing scenarios have been tried and/or suggested:

a) Irrigate every move across the block and then roll it back empty. This move sequence is similar to the way hand move lines were used. But a major disadvantage is the long distance the empty line is rolled back.

b) Skip irrigate every other move across the block and then on return irrigate the skipped sets on the way back instead of rolling back empty.

c) Irrigate every set across and then start, say twelve hours later, and irrigate every set coming back.

This avoids rolling the line empty all the way back, it also avoids some of the concern about the skip irrigation having alternating unirrigated spots on either side that may bother people. However, irrigating so soon after the previous irrigation causes extra deep percolation losses on the wet side of the field. There is also a longer than needed interval between irrigations on the opposite (or "dry") side. This mode is not recommended due to the alternate over irrigation followed by soil water shortage.

d) Irrigate every set across, roll the line empty for about three sets back and then irrigate the rest of the way back. Roll the line empty for three sets or so and then irrigate across again.

e) Other possible variations exist and depend upon soil, crop and field characteristics.

Maintenance Concerns

The wheelmove manufacturers provide adequate operation and maintenance instructions to fit their particular drive system. It is important that the owner/operator read and be familiar with and follow the instructions that match the equipment he/she is using. Check all nozzles and impact sprinklers for plugging, mismatched sizes, breakage, corrosion or other damage caused by wear or winter weather. Couplers and connections should be checked for leaks and repairs/replacements should be done early. It is a good practice to identify problem components at the end of the previous irrigation season and to have the replacement parts on hand for spring installation. If water leaks occur at joints or drain plugs during irrigation, check the gaskets and pipeline connections for wear or cracks and replace them as needed. Check and tighten the couplers and connectors as required.

In those places where an open ditch water supply is used, adequate screening and sediment removal are essential for minimum trouble operation. This is particularly a problem in the spring and early fall with more trash in the ditches and canals at those times.

There are quite a few anecdotal stories about runaway wheelmoves in high winds. Anytime the wheelmove is empty, and particularly during the off season, it is essential that it be anchored to prevent wind damage. Often this is as simple as moving the system over to a fence and tying it down in three or four places to sturdy posts. However, special wind anchors are available which attach to the lateral pipe and act like a brace to prevent movement.

Gravity Pressure Sprinklers

Many previously surface irrigated systems (private irrigation companies as well as individuals) have converted over to gravity pressure wheelmove irrigation. This switch from the traditional (often pioneer built) surface irrigation to gravity pressurised sprinkler has created some interesting situations. The value of "head end" water access opportunities in preference to a "tail end" location in multiple user surface irrigation systems is well known. However, when a conversion to gravity pressure sprinklers occurs, the historically "tail end" irrigators become switched to the "head end" position in that they now have the highest pressure. For some, this is the first time ever that they have had access to abundant water instead of the tail end dribble they are used to.

References

Hill, R.W., "Consumptive Use of Irrigated Crops in Utah", *Utah Agr. Exp. Stn. Res. Report No. 145*, Utah State University, Logan, UT. Oct., 1994.

_____________., "Utah Topography Favors Wheelmoves", *Irrigation Business and Technology*, November, 1997.

10

Tow Line Sprinkler Systems

A tow line sprinkler system consists of one or more sprinkler laterals made from 30 or 40 feet lengths of sprinkler tubing rigidly connected together. The laterals are towed from one set to the next with a tractor. Excluding the cost of the tractor, which is used for other farm functions, the tow line system is the most economical mechanical move system available and is convenient for pasture, hay and grain crops.

A tow line lateral usually consists of a quarter-mile length of 4-inch aluminium pipe in 40-feet sections rigidly coupled together. Three inch pipe has been used on short laterals and 30-feet lengths are sometimes used. Sprinklers on risers are located every 40. 60 or 80 feet along the lateral. Risers should be high enough that sprinklers will be above the mature crop. Automatic quick drain valves are located at each coupler to allow rapid drainage of water from the lateral. An empty line requires less force to pull to its next set reduces wear on the system and eliminates most cutting of ruts into the soil.

Two types of carriage are available. One is a skid plate attached at each coupler to slightly raise the pipe off the soil, protect the quick drain valve and provide a wear surface when towing the pipe. Two or three outriggers are required per quarter mile of lateral to keep the sprinklers upright. The other type uses small-metal wheels at 40-feet intervals along the lateral to raise the pipe and allow it to be towed easily.

Lateral Movement

The lateral is moved to a new position as follows: The supply line riser outlet or pump is turned off and the lateral is allowed to drain for 15 to 20 minutes. A time clock can be used to turn off the pump so the lines will be drained when the irrigator is ready to move them. The flexible coupler or hose connector to the mainline is disconnected. A tractor is connected to the end of the lateral nearest the mainline with a tow coupler and a pin or chain.

When towing the lateral, the driver guides the tractor across the mainline to a new position either one-half or a full set or lateral move away from its existing position on the mainline. After reaching the end of the field, the driver inserts a plug in the end of the lateral and returns to the mainline to attach the flexible coupler between the lateral and the mainline valve. The water now enters the lateral at the opposite end from the previous set.

$$\text{Usable water storage} \times \text{rooting depth} = \text{Total usable storage}$$

For example. if alfalfa is being raised on a sandy loam soil with 3 feet of soil over bedrock one obtains:

$$1.0 \text{ inch per foot} \times 3 \text{ feet} = 3.0 \text{ inches usable water storage}$$

If the usable water storage is now divided by the effective application rate, the time to fill the root zone is obtained:

$$\frac{\text{Usable water storage}}{\text{Effective application rate}} = \text{time to fill root zone}$$

For the example using an effective application rate of 0.22 inch per hour the time required to fill the root zone on the alfalfa crop would be 13.5 hours.

$$\frac{3.0 \text{ inches}}{0.22 \text{ inch per hour}} = 13.5 \text{ hours}$$

This would cause an inconvenient schedule. To work into the farming schedule moves are made on a 12 or 24 hour interval less the move time. In this case an 11.5 hour schedule would be used.

The amount of water applied per irrigation is equal to the effective application rate times the time per **set.**

Effective application rate x hours per irrigation = depth of water applied

In this example:

0.22 inch per hour x 11.5 hours = 2.5 inches.

The number of days between irrigations during the peak use period is obtained by dividing the amount of water applied per irrigation by the peak consumptive use per day for the crop at the farm location (Table 1).

$$\frac{\text{Water applied per irrigation}}{\text{peak use per day}} =$$

number of days between irrigations

If the farm in the example were located near Pocatello, the peak use of alfalfa would be 0.30 inch per day. Then the irrigation interval during this period would be 8 days.

$$\frac{\text{2.5 inches}}{\text{0.30 inch per day}} = \text{8 days}$$

The area irrigated per lateral can be determined by multiplying the area covered per set by the number of sets and dividing the total by the area of an acre. With a 1300-feet lateral and 60 foot moves, the lateral will irrigate 78,000 square feet per set. In 8 days, it will irrigate 16 sets or 16 x 78.000 — 1,248.000 square feet. Since an acre contains 43.560 square feet each lateral will irrigate 28.6 acres (1.248,000 ± 43,560 = 28.6).

Operational Problems

Tow line systems can operate over a wide range of surface topography. Sharp, cutup surfaces do require more outriggers to keep the system upright and increase the possibility of pipe breakage by causing greater strain on the pipe. However, tow lines have been operated satisfactorily on fields too rough for other mechanical move systems. More skid plates may be desirable under these conditions.

The tractor driver must make a gentle curve as he moves the lateral across the mainline toward the new position. Turning too short will kink the pipe a frequent type of damage to the system. Raising the end of the pipe with the tractor hitch helps to alleviate this problem.

While skid plates or wheel dollies support the pipe at regular intervals, the pipe slides over the soil surface between these points. This soil contact causes pipe wear especially on exposed sandy soils. Some systems are constructed to allow for pipe rotation to distribute wear around the pipe. The pipe should be rotated every 2 to 5 years, depending on manufacturer and soil conditions. One procedure is to rotate the pipe 180° so that the opposite side of the pipe touches the soil turn back a quarter at the next rotation then rotate 180° again. For maximum pipe life, follow the manufacturers recommendation on pipe rotation.

Frequently tow line systems are used on pastures where cattle are grazing. Cattle do not seem to bother operating laterals but they have damaged risers and sprinklers left in the field between irrigations. Loosening the bolts on the stabilising outriggers and laying the sprinklers on the ground will minimise this damage. The use of short risers, 1 foot or less, on pastures appears to reduce damage if the lines are left upright.

Attaching the tractor to the system usually requires driving onto wetted soil. With heavy crop cover and soils with good support, this may not pose a problem. Where traction or rutting is a problem, two alternatives can be used. Some operators plug or stop the first sprinkler, thus not irrigating the field along the mainline. While this solves the problem, some crop is lost in this area. Another solution is to attach a chain or cable to the line pull it out of the wet area and then attach the line directly to the tractor. This procedure requires a little more time but solves the problem without crop loss. The tractor wheels damage some of the crop. However, this is usually more than compensated for by labour saved in moving lines.

By using automatic quick drain valves and allowing 15 to 20 minutes drain time a small (20 to 30 horsepower) tractor can move the system. However, the temptation exists — especially with hired labour to use a large tractor at high speed, and to move the systems before drainage is completed. This greatly increases the possibility of pulling the lateral apart, particularly where there is a chance of catching skids or wheels in rocky areas.

Tow lines have worked well in rocky fields after rocks are cleared from the tow path. Rock outcroppings near the mainline present problems if the pipe must be moved diagonally across them. Either the outcroppings must be levelled or the pipe must be lifted over them.

System Design Characteristics

Water Supply Requirements

The design of a tow line system depends on the water use by the crops to be grown, Table 1 lists the average peak crop water use for alfalfa in inches per day for different areas of Idaho and the design water requirements in gallons per minute per acre for two design efficiencies. Since alfalfa is one of the crops most frequently grown under tow lines, it may be used for design. The peak water use of grain and pasture crops is about 10% less than that of alfalfa. Most systems can be designed using the 70% figure if winds are not excessive. For every mile per hour of average wind speed over 5 miles per hour the design efficiency should be lowered 1%. Thus, with an average 10-mileper-hour wind, the design efficiency should be 65%.

The design water requirement listed in Table 1 should provide sufficient water to meet peak water requirements in most years. Where insufficient water is available, the system may be designed for the water available with the expectation of reduced yield. Another solution would be to raise crops whose periods of peak requirements do not coincide. For example grain may be grown under the same irrigation system with alfalfa and pasture. As the grain crop approaches maturity, the tow lines may be moved to the alfalfa and the pasture to meet their peak requirements.

Table 1. Peak water requirement for alfalfa at different Idaho locations and the corresponding design water requirements in gallo per minute per acre at two application efficiencies.

		Design water requirement	
Area	*Peak water requirement*	*70% efficiency*	*65% efficiency*
	(inches/day)		(gallons/minute/acre)
Banners Ferry	0.25	6.7	7.2
Caldwell	0.33	8.9	9.6
Council	0.29	7.8	8.4
Dubois	0.28	7.5	8.1
Grand View	0.34	9.2	9.9
Grangeville	0.29	7.8	8.4
Idaho Falls	0.29	7.8	8.4
Lewiston	0.34	9.2	9.9
Moscow	0.29	7.8	8.4
Pocatello	0.30	8.1	8.7
Twin Falls	0.33	8.9	9.6

A farm completely devoted to pasture and alfalfa will require a larger peak flow of water than farms with mixed crops in the same area.

Mainline Design

The mainline should be laid in a ditch or completely buried to facilitate towing the lateral across it. Mainline risers should be installed in line with lateral sets. Then the lateral would be towed across the mainline between risers.

Occasionally tow lines are installed on existing mainline systems at different spacings than the riser outlets. In this case short posts or similar guards which do not catch the outriggers should be placed near the riser outlet to avoid damage caused by contact between the towed lateral and mainline riser. A portable mainline may be installed in a trench deep enough to place the risers below the soil surface to avoid scraping the lateral on the riser.

Lateral Design

When possible the mainline should be laid across the center of the field. The laterals will then be half the length of the field. On fields with nonparallel sides a few lengths of hand line can be added to the tow line to complete the field, but adding hand lines detracts from the mechanical move feature of the tow line. One system uses a quick release coupler clamp to join pipe lengths. This allows the irrigator to easily drop lengths of pipe when irrigating nonrectangular fields.

The selection of sprinkler spacing on the lateral distance between lateral sets sprinkler type pressure and nozzle size are normally determined by a combination of economic climatic and soil factors. With average windspeed less than 5 miles per hour sprinkler manufacturers recommend that the maximum spacing of sprinklers along the lateral not exceed 40% of the diameter of coverage of the sprinkler and 65% of the diameter between laterals.

Medium-size agricultural sprinklers with 5/32, 11/64 or 3 /16-inch nozzles, operated at their recommended pressure of 50 to 60 psi, have a diameter of coverage from 90 to 100 feet and are usually used on 40 by 60 feet spacings on tow lines.

Larger agricultural sprinklers with 7/32, 1/4 or 9/32-inch nozzles operated at recommended pressures of 60 to 70 psi have a diameter of coverage from 130 to 150 feet and should be used on a 60 by 80 feet spacing. These sprinklers are sometimes used at 80 by 80 feet and 80 by 90 feet spacings, although these exceed the manufacturers' suggested limits and are accompanied by poorer water distribution and yield reductions.

Generally, it is advisable to purchase a pump and mainline that are large enough to slightly exceed these recommended pressures so that system wear does not cause reduced water application efficiency in the future.

The choice between 60 and 80 foot moves involves a trade-off between smaller nozzles which require more laterals per farm, more moves per season and more tow paths as opposed to larger nozzles which require more pressure with an accompanying larger motor and increased power costs. Estimates of the comparative costs of irrigating the farm with the two spacings for one season will help in .making the decision.

Sprinkler Size Selection

After the selection of sprinkler spacing and pressure a nozzle size is selected that will apply water at a rate equal to or less than the infiltration rate of the soil. If the infiltration rate is exceeded, water will run off the slopes leaving dry areas and creating wet spots in low places. The results will be decreased crop production fertiliser leaching in low spots and possible soil erosion.

The sprinkler application rate at 100% efficiency can be obtained from sprinkler manufacturer's tables if the nozzle size pressure and spacing are known. This value is multiplied by the application efficiency for example 70% for low wind conditions) to arrive at the effective application rate.

Application rate at 100 % application efficiency
effective x application rate

As an example consider a 40 x 60 feet spacing with 3/16-inch nozzles operated at 60 psi. The application rate at 100% efficiency would be 0.32 inch per hour. The effective application rate would be:

0.32 inch per hour x 0.70 = 0.22 inch per hour.

The extra 0.10 inch per hour falls as spotty excess water due to sprinkler patterns.

The usable storage capacity depends on the crop its rooting depth and the water holding capacity of the soil. Potatoes, beans and some other row crops are sensitive to water stress and yield can be reduced when more than 40% of the available water is depleted. Grain, pasture and alfalfa are more tolerant to water stress and can use 60% of the available water before production is seriously affected.

Table 2. Available water at field capacity and usable water storage for alfalfa and grain in soils of different textures.

Soil texture	*Total available water at field capacity in inches per foot of soil*	*Usable water storage in inches per foot of soil*
Sandy	0.8	0.5
Loamy sand	1.0	0.6
Sandy loam	1.7	1.0
Loam	2.0	1.2
Sift loam	2.2	1.3

Table 2 indicates the approximate total water-holding capacity per foot of soil for various soil textures. The right-hand column indicates the approximate usable storage per foot of soil for hay and grain crops and can be used to estimate the amount of water available to the crop. The value in the right-hand column for the soil texture in the field must be multiplied by the rooting depth of the crop to obtain the total inches of readily available water in soil storage. A recommended practice when using 12-hour sets is to use an odd number of lateral sets in returning to the same location. Thus alternate sprinklings at the same location will be made during the day one time and at night the next irrigation. Because winds are usually less at night, this practice increases the uniformity of cumulative applications and helps prevent the development of wet or dry spots.

The movement pattern used to return a system to its original position most conveniently consists of skipping every other set and filling them in on the trip back to the starting position. This pattern requires an even number of sets.

Cost

Quarter-mile tow line laterals complete with flexible couplers sprinklers and end plugs cost about 165 to 170 percent as much as a hand line and 46 to 48 percent as much as a sideroll or wheel move lateral, If a hand line cost $1,000, a tow line would cost about $1,650 to $1,700 and a sideroll about $3,500. This does not include the cost of the pump and motor, water and mainline with riser outlets. If a tow line system with 80-foot moves between lateral sets is compared to the other systems with 60-foot moves, a larger pump and motor will be needed to increase line pressure 15 to 20 psi.

A good farm manager will consider the annual overhead costs of depreciation, interest on investment and insurance together with the annual operating costs of labour, power maintenance and interest on operating capital.

The disadvantage of the tow line system is the damage done to row crops in the turning area. As a result, turning areas are usually planted to alfalfa. Tow lines work best in rectangular fields. The major advantages of tow lines are their low cost for a mechanical move system and their ease and speed of movement.

References

Small L E & Carruthers I D., *Farmer-financed irrigation: the economics of reform*, Cambridge University Press, Cambridge, UK, 1991.

Vermillion D L., "Impacts of irrigation management transfer: a review of the experience", *Research Report No.11*, IIMI, Colombo, Sri Lanka, 1997.

11

Irrigation Scheduling Methods

Irrigation scheduling concerns the farmers' decision process concerning 'when' to irrigate and 'how much' water to apply in order to maximise profit. This requires knowledge on crop water requirements and yield responses to water, the constraints specific to each irrigation method and irrigation equipment, the limitations relative to the water supply system and the financial and economic implications of the irrigation practice. Thus, the consideration of all these aspects makes irrigation scheduling a very complex decision making process, one which only very few farmers can understand and therefore adopt.

Research has made available a large number of tools including procedures to compute crop water requirements, to simulate soil water balance, to estimate the impact of water deficits on yields and to estimate the economic returns of irrigation. Despite this vast number of tools of varied nature, irrigation scheduling is not yet utilised by the majority of farmers. Furthermore, only limited irrigation scheduling information is utilised worldwide by irrigation system managers, extensionists or farmer advisers. It is recognised, however, that the adoption of appropriate irrigation scheduling practices could lead to increased yields and greater profit for farmers, significant water savings, reduced environmental impact of irrigation and improved sustainability of irrigated agriculture.

Consequently, there is a need to better identify the factors that could enhance the adoption of appropriate irrigation scheduling practices, favour the transfer of technology from research to practice, and give new orientation to researchers. These aspects are particularly important in that they concern the inter-relationship between on-farm irrigation systems and irrigation scheduling, and involve two related disciplines: irrigation engineering and agronomy.

Scheduling Irrigation under Different Irrigation Methods

Surface Irrigation Systems

Surface irrigation is the most common irrigation method worldwide. There are several types of surface systems, namely:

— *Basin irrigation:* Water is applied to levelled surface units (basins) which have complete perimeter dikes to prevent runoff and to allow infiltration after cut-off. The best performance is obtained when advance time is minimised by using large non-erosive discharges and the basin surface is precision levelled. This method is the most commonly practised worldwide both for rice and other field crops including orchard tree crops. In general, basins are small and water application is manually controlled. Basin irrigation using large laser-levelled units with automated or semi-automated control is practised in a few areas in developed countries.

— *Furrow irrigation:* Water is applied to furrows using small discharges to favour water infiltration while advancing down the field. Fields must have a mild slope, and inflow discharges must be such that the advance is not too fast, producing excessive runoff losses, nor too slow, which induces excessive infiltration in the upper part of the field. Short blocked furrows with manually controlled water applications are practised by traditional irrigators. Long and precisely levelled furrows with automated or semi-automated control have become increasingly popular in developed countries.

— *Border irrigation:* Water is applied to short or long strips of land diked on both sides and open at the downstream end. Irrigation takes place by allowing the flow to advance and infiltrate along the border. Larger inflow rates are utilised when the field slope is very small with management becoming similar to that of basins. Similarly, basin irrigation automated water control is often applied.

There are other surface methods like contour ditch, wild flooding and water spreading, but these are of less significance. In general, these types of surface irrigation systems are manually operated and often the modernisation of irrigation practices leads to their conversion to other types of systems.

Two types of systems can be considered under the perspective of irrigation scheduling practices:

— *Traditional systems:* The water control is carried out manually according to the ability of the irrigator. In small basins or borders and in short furrows, a common practice is where the irrigator cuts off the supply when the advance is completed. Because the fields are often uneven and the discharges are variable and largely unknown, this practice induces large variations in the volumes of water applied in each irrigation. It is therefore extremely difficult to control 'how much' water is applied.

— *Modernised systems:* In these systems, some form of control of discharge such as syphons, gated pipes, lay-flat tubes or gates, and/or some form of automation including surge flow valves, cablegation or automated canal gates is used. In this case the fields are often levelled, while the advance and supply times and the inflow rate can be measured or estimated. Thus, in these systems, it is possible to control 'how much' water should be applied at a given irrigation.

Relationship between Irrigation Performance and Irrigation Scheduling

The two main parameters used to characterise surface irrigation performance are the distribution uniformity and the application efficiency. The distribution uniformity DU (%) is defined as the ratio of the average infiltrated depth in the low quarter of the field (Z_{lq}, mm) to the average infiltrated depth. (Z_{av}, mm) as follows:

$$DU = 100 \frac{Z_{lq}}{Z_{av}} \tag{1}$$

The distribution uniformity results from several design and management variables that characterise an irrigation event. Functionally, it can be described as:

$$DU = f_1 (q_{in}, L, n, S_o, I_c, F_a, t_{co}), \tag{2}$$

where:

q_{in} inflow rate to the furrow (or per unit width of the border or basin)

L length of the furrow (or border, or basin)
n roughness coefficient

S_0 longitudinal slope of the field

I_c intake characteristics of the soil

F_a furrow (or border, or basin) form

t_{co} time of cut-off

The application efficiency e_a (%) is defined as the ratio of the average depth added to the root zone storage (Z_r, mm) and the average depth applied to the field (D, mm), as follows:

$$e_a = \frac{Z_r}{D} 100 \tag{3}$$

The application efficiency also depends on several design and management variables. This can be expressed functionally as follows:

$$e_a = f_2 \, (q_f\,,\ L,\ n,\ S_0 I_c,\ F_a,\ t_{co},\ SMD), \tag{4}$$

where SMD is the soil moisture deficit at the time of irrigation. All other variables are as defined above. It can be seen from Equations 2 and 4 that the application efficiency depends on both the distribution uniformity and the irrigation scheduling. The latter controls the value of SMD.

Field evaluations of irrigation systems play a fundamental role in improving surface irrigation management. They provide the information required for design, model validation and for advising irrigators on how to improve their management practices. The field observations required for conducting a system evaluation include: inflow and outflow volumes; irrigation times/phases, particularly advance and recession; soil moisture before and after irrigation; the slope/topography of the field being irrigated; and the management rules currently utilised by the irrigator.

Field Implementation of Irrigation Scheduling

Most surface irrigated areas are supplied from collective irrigation canal systems. Farm irrigation scheduling depends upon the delivery schedule, e.g., rate, duration and frequency of irrigation are dictated by the system's operational policies. Discharge and duration impose constraints on the volume of application and frequency determines the irrigation timing. In general, surface irrigation delivery systems are rigid and the time interval between successive deliveries is often too long. Irrigators tend to compensate for this by applying all the water they are entitled to use. For deep rooted crops and soils without excessive permeability this strategy may be appropriate. However, for shallow rooted crops and/or soils with small water holding capacity, crops may become highly stressed and percolation losses can be substantial.

The realisation of the benefits of modernisation of surface irrigation are constrained by rigid delivery schedules. Irrigation scheduling can only be applied if farmers are in control of supply, or when the delivery schedule is sufficiently flexible to adequately select 'when' to irrigate. This applies to both traditional and modern irrigation. In the case of modernised surface systems, the adoption of improved irrigation scheduling methods should go, as it often does, together with the improved operation of the delivery system to allow more flexibility in selecting the appropriate inflow rates and supply time.

Substantial improvement can be achieved if delivery is made by an arranged delivery schedule. This normally requires a certain notice time between the demand for water by the farmer and the time of supply. Often, this can be achieved more easily when users are in control of the distribution units. Farm surface irrigation systems that rely on large irrigation depths and large irrigation intervals have enough flexibility to accommodate a demand lag time of two or three days without inducing severe water stress for most crops. This could favour the adoption of arranged delivery schedules.

In general, it is difficult to apply small irrigation depths by surface irrigation. Smaller depths can be applied in basin irrigation provided that the land is precision levelled and the discharge rate is sufficiently large to obtain a fast advance. In furrow irrigation the time of cut-off should be larger than the advance time to provide for sufficient intake opportunity time; while in border irrigation, inflow is normally cut off before the advance front has reached the end of the field. This implies that surface irrigation applications are usually large which makes this method more suitable for situations where the SMD and the management allowable deficit (MAD) are relatively large. Under these circumstances scheduling can be simplified as proposed by Hill and Allen.

The adoption of simple irrigation scheduling calendars is advantageous for collective system management and for introducing improved tools for controlling the discharge rates at farm level. It is obvious that, to achieve this, it is necessary that volume delivered be known, e.g., irrigators should have control on the discharge rate and duration. Another irrigation scheduling option is to use irrigation scheduling simulation models to perform the soil water balance using soil, crop and meteorological data and to provide information on the soil moisture status. However, the use of models present serious difficulties for application in real time since they require not only validation but also a support advisory service and training of irrigation extensionists, managers and/or farmers.

The use of soil water monitoring devices is generally more appropriate for large fields and crops that may economically justify the costs of purchase and operation. Normally, the use of these devices requires the selection of an irrigation threshold corresponding to the MAD. Observations of the SMD have to be regularly performed in time and space to enable the prediction of the irrigation date when SMD = MAD. A further degree of sophistication can be achieved by combining soil water balance simulation models with soil moisture observations. This can be done in two ways: (a) soil moisture information is used to validate the model predictions, and (b) continuous soil moisture and meteorological monitoring coupled with a model for forward prediction of the soil water status.

Sprinkler Irrigation systems

Sprinkler irrigation systems are normally used under more favourable operational conditions than surface systems as farmers can better control the discharge rates, duration and frequency. Many sprinkler systems have independent water supply or are connected to networks which may be operated on demand. However, the pressure head from the hydrants is often not appropriate resulting in lower (or higher) discharges than those envisaged during the design phase. Pressure head variations at the hydrant should be identified by the user if appropriate equipment is available.

Different types of sprinkler systems can be considered in relation to irrigation scheduling:

— *Set systems:* Sprinklers irrigate in a fixed position. There are no limitations on the duration of the set time. Thus, they can be utilised to apply small depths and frequent waterings (not suitable for the movable systems because of operational constraints), or large depths and large irrigation intervals (less appropriate for systems with very low application rates).

— *Travelling guns, hose-reel or hose-pull:* These systems in general apply irrigation depths ranging from 15 to 40 mm. These, systems are not suitable for applying very small or large depths, mainly because of limitations in system capacity;

— *Continuous move laterals:* These systems are well adapted to apply small and frequent irrigations.

The hydraulic design of sprinkler irrigation is generally well understood. The overall sprinkler design process is, however, more complicated since it requires the selection and the best combination of the following: pressure head (H), nozzle diameter (D_n), sprinkler discharge (q_s) and wetted diameter (D_w); spacings between sprinklers and/or rows which depend on the application rate (i_s), wind conditions and distance between rows of tree crops; and sprinkler characteristics including the size of drops and the distribution pattern, in relation to soil and climate conditions.

The most common design criteria for sprinkler laterals is that sprinkler discharge (q_s) should not vary by more than 10% between the points of highest and lowest pressure in the system. However, this condition is often not achieved in practice. Augier *et al.* report that in France 56% of solid set systems evaluated showed a pressure variation higher than 20%.

Many sprinkler systems have not been properly designed, are not operated according to design rules, or their operation is hampered by poor maintenance. This results in inadequate pressures and discharges along the system. Unfortunately, these problems are much more frequent than expected even in developed countries: in France, as reported by Augier *et al.*, the actual application depths deviate by more than 20% from the design ones in 46% of travelling guns and 34% of solid set systems. This would indicate that the main limitation to appropriate scheduling is that most farmers do not know the discharges delivered by their systems and are therefore only in control of the duration.

In case of direct supply by a farm pumping system, the pressure head can also be lower (or higher) than expected due both to design shortcomings or lack of maintenance. When appropriate measurement devices are installed, these problems can be identified and either the set time or the travelling speed can be corrected according to the actual discharge.

Relationship between Irrigation Performance and Irrigation Scheduling

Christiansen's coefficient of uniformity CU (%) is the most commonly used performance indicator to index application uniformity. It is defined as:

$$CU = 100\left(1 - \frac{\Sigma X}{nm}\right) \quad (5)$$

where

X absolute deviations of application depths observations from the mean (millimetres) m mean of observed depths (millimetres) n number of observations.

The coefficient of uniformity is determined by several design variables and can be functionally expressed by the following relation:

$$CU = f_{1\_}\ (H,\ D\ H,\ S,\ D_{n},\ DP,\ W), \quad \textbf{(6)}$$

where

H pressure head available at the sprinkler

Δ H variation of the pressure in the operating set, or along the moving lateral

S spacing of the sprinklers along the lateral (and between laterals) or spacings between travellers

D_n nozzle diameter, which influences the discharge q_s and wetted diameter D_w for a given pressure head

DP the distribution pattern of the sprinkler

W wind speed.

All the above variables are set during at the design stage. However, spacings are often oversized. Augier *et al.* report that excessive spacing occurs in 65% of travelling guns and 70% of solid set systems evaluated in France.

The application efficiency e_a (Equation 2) depends not only on the design but on management variables and can be functionally described by:

$$e_a = f_2 \ (H, \Delta H, S, D_n, DP, W, I_c, i_s, t_i, SMD), \qquad (7)$$

where

I_c intake characteristics of the soil

i_s application rate of the sprinkler

t_i duration of the irrigation event

SMD soil moisture deficit before the irrigation event.

Most of the parameters in Equation 7 are controlled by the designer, while the duration of the application and SMD are management variables controlled by the irrigator. Hence, as with surface irrigation, the application efficiency depends upon the application uniformity and the irrigation scheduling. Field evaluations of systems under operation are very useful for improving management, including irrigation scheduling.

Field Implementation of Irrigation Scheduling

Overall, in sprinkler irrigation, the management of the system is easier than in surface irrigation and performance is largely dependent on the system design. Therefore quality control of services, equipment and design are very important in supporting farmers' operations.

For set systems and travelling guns, the MAD is commonly smaller than for surface irrigation. The methods for irrigation scheduling that can be used are the same as for surface irrigation: (a) soil water balance simulation models; (b) monitoring the soil water status; and (c) a combination of both. The observation of the SMD requires more accuracy and to be more frequent when irrigation depths are small. In areas with a large number of small farmers the use of simple irrigation scheduling calendars may be more appropriate.

Irrigation scheduling models for sprinkler irrigation to serve several farms may provide not only information on when and how much to irrigate but also the operation plan for the individual irrigation machines. When frequent irrigations are applied the error in estimating the SMD may be larger than the depths to be applied. This is the case for daily irrigations with centre pivot laterals. Thus, other scheduling strategies based on the replacement of the volume of water consumed during the precedent irrigation interval must be applied. For centre pivots, special irrigation scheduling models have been developed which associate modules for water application, fertigation, chemigation and energy management.

Trickle Irrigation

Trickle irrigation systems are set systems that use a variety of low discharge emitters to apply water to the field. Systems for tree crops often have one or more emitters per tree depending on the wetted area, hydraulic properties of the soil, emitters discharge (q_e) and emitters wetting pattern.

There are several types of emitters, including drippers, sprayers and micro sprinklers, microjets, and bubblers.

For row crops, a lateral supplies water for every one or two crop rows when drippers are utilized. Underground perforated pipes have also become popular. When the soil is highly permeable sprayers and micro sprinklers are preferable.

Emitters discharge q_e (litres per hour) relates to the pressure head H (metres) through the following relation:

$$q_e = K H^x, \quad (8)$$

where

x emitter discharge exponent

K discharge coefficient of the emitter.

The parameter x varies from near x = 0, for self compensating emitters, up to x = 1, for laminar flow emitters. Most commonly, x values vary from 0.4 to 0.55. Self-compensating emitters are used to minimise discharge variations along very long laterals or in sloping laterals.

The design of trickle systems involves the selection of the appropriate combination of: pressure H, discharge q_e and emitter discharge exponent x; percent area wetted; emitter type and spacings; emitter susceptibility to clogging, cleaning and ability to undergo variations in temperature; type of filters, their location and cleaning capability; requirements for application of fertilisers and chemicals, control of pressure and/or flow, type of devices and their location in the system; and automation facilities.

To achieve this, the designer must know not only the irrigation requirement of the crop but also the duration and frequency of irrigations (including fertigation/chemigation requirements). Hence, design and irrigation scheduling processes are intertwined. Performance is highly dependent on the quality of design and equipment selected.

Relationship between irrigation performance and irrigation scheduling

The most useful system performance indicator for trickle systems is the emission uniformity EU (%) which, in the case of field evaluation, is defined as:

$$EU = 100\ q_{lq}/q_n, \quad (9)$$

where

q_{lq} average discharge rate of low quarter of emitter discharge observations (litres/hour)

q_a average discharge rate of all observations (litres/hour). For design purposes, EU is defined as:

$$EU = 100\left(1 - 1.27\frac{C_V}{\sqrt{N}}\right)\frac{q_n}{q_a} \tag{10}$$

where

C_v coefficient of manufacturing variation for the emitter

N number of emitters per tree (N = 1 for row crops)

q_n minimum emission rate corresponding to minimum pressure in the system (litres/hour)

q_a average (or design) emission rate (litres per hour).

The emission uniformity is determined by a combination of design parameters which can be functionally expressed as follows:

$$EU = f_1''\ (H,\ \Delta H,\ x,\ E_c,\ C_v,\ FI), \tag{11}$$

where

H pressure head at emitters

ΔH variation in pressure along the system

x emitter discharge exponent

E_c emitter characteristics related to variations in discharge

C_v oefficient of manufacturing variation for the emitter

FI filtering capabilities of the system

The above-mentioned parameters mostly depend on design decisions. Thus, as for sprinkler systems, the irrigator has very little control on the uniformity performance of the system except for the maintenance of the system.

As with surface and sprinkler irrigation systems, the application efficiency e_a is also determined by the combination of design and management variables as indicated in the following functional relation:

$$e_a = f_2'' (H, \Delta H, x, E_c, C_v, FI, K_s, SW, t_i, D\ t_i), \qquad (12)$$

where

K_s hydraulic conductivity of the soil

SW soil water conditions before irrigation

t_i duration of the irrigation

D t_i time intervals between irrigations

The irrigator's control is basically reduced to the timing of irrigation and the volumes and frequency of applications. Thus to achieve good performance in trickle irrigation good support to farmers in terms of quality control of equipment, services and design is required. Field evaluations of systems under operation can play a fundamental role in improving performance.

Field Implementation of Irrigation Scheduling

Farmers adopting trickle systems are in control of flow rate, duration and frequency of irrigations. However, as with sprinkler systems, they may not be aware of the discharge when pressure head variations occur in the supply system or when systems are not properly designed (these problems can be reduced when self-compensating emitters are utilised).

Most of the problems derive from inappropriate design and selection of equipment. It is often said that a localised irrigation system is selected or should be selected to save water when in fact such a system should be selected to maximise yields and profits. For these systems, management is tied up with design. The decisions made during the design phase may affect to a large degree the operation and management of the system.

Trickle irrigation systems are normally used for frequent or very frequent applications. Under intensive horticultural production more than one application per day may be necessary. Larger irrigation intervals of a few days are appropriate for deep rooted orchard trees.

Irrigation scheduling can be based on crop indicators or soil water status. Videotel or Irritel systems, as well as other forms of water balance models are also appropriate. For very frequent waterings the irrigation and fertilisation strategies must be integrated.

Environmental Impact and Control

The most common environmental impacts of on-farm irrigation are:

- — soil salinisation due to use of saline water or to the rise of saline water table;
- — waterlogging due to excess water application;
- — nitrate and pesticide contamination of the groundwater (and surface waters) due to excessive use of chemicals in intensive agricultural production and/ or to overirrigation;
- — soil erosion due to surface runoff from surface and sprinkler irrigation systems;
- — soil degradation due to modifications of the soil profile from inappropriate land grading;
- — deterioration of the soil structure from both surface flow and sprinkler raindrops;
- — degradation of water bodies receiving saline irrigation return flows.

These impacts can be reduced when appropriate water application techniques coupled with irrigation scheduling are applied. In this regard, it is, however, necessary to stress the following aspects:

- — Irrigation scheduling techniques cannot be properly implemented nor produce the desirable consequences unless proper water application practices are implemented. This may be one reason why innovations in irrigation scheduling techniques are not widely adopted in practice;
- — Application efficiency depends not only on system design criteria but also on the volume and timeliness

of water applications, e.g., scheduling decisions. Thus, the adoption of improved hardware is not by itself enough to attain high levels of performance.

High application uniformity minimises the differences in water application over a field, e.g., areas receiving too much or too little water. Application uniformity affects both yields and the extent of environmental impact from irrigation.

References

Buchleiter, G.W., "Improved irrigation management under center pivots with SCHED", In: *Crop -Water- Simulation Models in Practice*, L.S. Pereira, B.J. van den Broek, P. Kabat and R.G. Allen (eds.). Wageningen Pers, Wageningen, 1995.

Burt, C.M., "Essential water delivery policies for modem on-farm irrigation management", In: *Irrigation Scheduling: From Theory to Practice, Proceedings ICID/FAO Workshop*, Rome, Sept. 1995.

Goussard, J., "Interaction between water delivery and irrigation scheduling", In: *Irrigation Scheduling: From Theory to Practice, Proceedings ICID/FAO Workshop*, Rome, Sept. 1995.

Hoffman, G.J., Howell, T.A. and Solomon, K.H. (eds.), "Management of Farm Irrigation Systems", *ASAE*, St. Joseph, 1990.

Keller, J. and Bliesner, R.D., *Sprinkler and Trickle Irrigation*, Van Nostrand Reinhold, New York, 1990.

Mao, Z., "Environmental impact of water-saving irrigation for rice", In: *Irrigation Scheduling: From Theory to Practice, Proceedings ICID/FAO Workshop*, Rome, Sept. 1995.

Walker, W.R. and Skogerboe, G.V., *Surface Irrigation, Theory and Practice*, Prentice Hall, Englewood Cliffs, New Iersey, 1987.

12

Irrigation System Maintenance

Irrigation provides some 40 % of the world's food from only 17 per cent of the global cropped area. It is estimated that 2 400 million people depend on irrigated agriculture for food and livelihood. The high level of investment by multilateral development banks and national governments indicates the importance attached to irrigation by funders. In the period 1980-90, the World Bank and the Asian Development Bank were together lending some US$1 500 million annually for irrigation. Lending by the Banks supported national spending programmes. In the early 1980s, India was investing some US$2 400 million in irrigation annually.

Although irrigated agriculture still receives a large amount of funding, expenditure over the last decade has fallen significantly due to the following factors:

— World food prices are close to historic lows, reinforcing a perception of food sufficiency which developed when global grain stocks stood at over 100 days supply, following the Green Revolution in the 1960s. Stocks are now declining, currently standing at 45 days supply.

— Donors now perceive the economics of irrigation development as less favourable than hitherto. The unit cost of development increased as the most suitable areas were exploited, while food prices were falling. Nonetheless, a review of 585 irrigation projects found that the average economic internal rate of return (EIRR) was 15 %, substantially above the 10 %

opportunity cost of capital. Projects in the irrigated agriculture subsector generally perform as well as other types of investment.

— Most investment is now for the rehabilitation of existing schemes. The per hectare cost of rehabilitation is considerably less than for a new project. When conventional economics writes off the cost of the original work, rehabilitation appears an attractive option.

— Competition for available water supplies among different users is growing. Many planners view irrigation as a low value, low efficiency and highly subsidised use of water.

— Environmental and social issues, involving actual or potential problems with resettlement, human health or land degradation, have delayed, modified or, in some cases, halted irrigation developments.

Though the rate of increase of the global population now seems to be slowing from the more pessimistic estimates, irrigation must play an increasingly important role in sustaining the world's food supply. FAO estimated that 60 per cent of the extra food required must in future come from irrigation. It is therefore essential to continue to invest in irrigation, and to safeguard existing investments. In the face of increasing competition from other water users, irrigated agriculture must make better use of a diminishing resource. In common with programmes to reduce leakage from potable water supply networks, effective and timely maintenance on irrigation schemes can improve the efficiency of water distribution and help to manage demand.

Consequences of Poor Maintenance

Many irrigation systems constructed within recent decades are failing within their design lifetime. About two-thirds of recent international lending has been for systems which have suffered premature technical failure. Many millions of dollars invested in the original infrastructure are being written off and anticipated returns have not been achieved because maintenance is inadequate.

Some argue that a high level of maintenance leads to postponement of the date at which a system should really be modernised. It is contended here that, if major changes to the system design appear necessary after as little as 15 years, then the original design brief was inadequate. Such issues are not further considered here except to note that, in a context where maintenance practices are poor, the implication that underfunding should be accepted, with the objective of undertaking early modernisation, appears to send the wrong signals to maintenance agencies. To endorse prevailing cycles of *build-neglect-rebuild*, and to institutionalise unplanned obsolescence, would seem to conflict with the concept of sustainability.

Schemes in arid and semi-arid areas, where crops are entirely dependent on irrigation under managed rotations, are particularly at risk from poor maintenance. In the humid tropics, schemes growing paddy rice under continuous irrigation, supplementary to substantial monsoon rainfall, are inherently less sensitive to maintenance defects.

Conventional economics, using a high discount rate for future costs and benefits, fails to show the importance of maintenance in sustaining the life of a system and the livelihoods of farmers. Capital, often readily available from international sources, is taken as the scarce resource, whereas in practice it is the recurrent funding from local sources that is scarce. Tiffen points out that since costs and benefits occurring in the future are highly discounted, little benefit is apparently to be derived from extending the life of a new project beyond 10-15 years. The result of this thinking is that a project with a low initial cost, which deteriorates quickly and is dependent for continued survival on timely and properly funded maintenance, is preferred to one that is constructed to need less maintenance, because it appears cheaper. Yet for a farmer, and also for a nation, it is important that a scheme endures. Tiffen concludes that overall economic criteria should be supplemented by calculations of the costs and benefits to farmers, and to project administration, to ensure that the costs of O&M can be covered.

The World Bank has frequently expressed concern that projects may not be sustained, and that governments must face increased recurrent costs. Yet Price comments: "... the idea of sustainability, by giving absolute importance to future generations, is the ideological opposite of discounting ...". Others – Finney and Carruthers – have made efforts to modify conventional analysis to overcome the economic bias against sustainable development but, in the absence of consensus, economic justification for projects continues to follow existing methods.

Carruthers and Morrison summarised the effects of poor O&M on project performance, as follows:

- systems work below capacity, erratic water supplies reduce the cultivated area and depress yield;
- farmers shift to lower value crops, to reduce risk, limiting the use of inputs and reducing investment.

There are also longer-term consequences:

- premature rehabilitation results in a major loss of production while systems are shut down for several cropping seasons;
- international loans for rehabilitation need to be repaid in scarce foreign currency earned from exports;
- local funds for rehabilitation are not available to other sectors of the economy, so important programmes are postponed;
- loss of productive capacity may require food imports, which must be paid for in foreign currencies.

The resulting cycles of *build-neglect-rebuild* have had serious social and financial implications for farmers. Those in the more favoured parts of the system may continue to receive an adequate water supply, while those at the tail-end can face ruin. Poor maintenance thus directly aggravates existing inequities within the farming community. It initiates a vicious circle of decline: reduced water supply; lost output; farmers'anger, despair and reduced investment; reduced water fee collections; vandalism and conflict. Smallholder farmers operate on a narrow margin between relative success and failure. Land is often mortgaged against the following harvest to pay for the cost of agricultural inputs. A single disastrous cropping season can mean the loss of a farmer's land and enforced migration to the cities to seek work.

Poor maintenance can also produce adverse consequences for the environment, such as waterlogging and salinity resulting from impeded drainage, and for health, through the increased incidence of water-related diseases associated with blocked channels and stagnant waters.

Considering that farmers are the principal victims of poor maintenance, it may seem illogical that they commonly withhold payments for O&M, thus aggravating the problem of underfunding. However, in many countries farmers' experience over many years is that governments have implicitly accepted declining O&M payments, and that sometimes politicians have actively courted support by promising reduced fees. The O&M system is now widely seen as entirely the responsibility of government, a public good. The stark fact that, in the current financial climate, developing country governments are no longer able to shoulder the responsibility unaided has yet to be appreciated.

There are a number of reasons why system maintenance has been widely underfunded by governments:

— *Maintenance has a low priority.* The physical effects of underfunding do not appear for several years beyond the horizon of financial planners and the economic implications are not understood. Functional failures are not dramatic and publicised. Irrigation system staff can operate canals for a while by infringing on freeboard3 and cropping systems show some resilience. So projects may appear to function adequately for a time, even though they are deteriorating and action to stem decline is therefore postponed.

— Irrigation departments must justify their budgets to the finance ministry, which views maintenance as a net drain on national resources. The returns from water fees – commonly collected by the revenue department and remitted direct to central government – fall far short of the costs, owing to inflation and poor rates of collection. More profitable areas of the economy invariably take precedence in the competition for scarce resources. The sum ultimately allocated for O&M is unrelated to the amount collected in water fees. There are thus no incentives for an agency to improve its performance which is, in the absence of adequate monitoring arrangements, effectively unquantifiable.

— The international lending agencies are reluctant to fund recurrent costs owing to concern about lack of accountability, and a belief that support for O&M will merely postpone the time when nations must assume responsibility for their own projects.

For example in India, underfunding of O&M costs has led to a situation where some US$2 300/ha (1988/89 prices) is spent on the development of irrigation facilities, whereas existing irrigation potential is underutilised for the lack of US$20/ha for maintenance. Without funds to meet this small recurring cost the productivity of the entire system which has been built up at enormous cost, will fall to abysmally low levels. Shortage of funds is one reason for inadequate maintenance, but it is very far from being the only cause. The difficulties are compounded by other problems:

— Maintenance suffers by being grouped with Operations within the O&M budget. Available money is generally allocated in the following order: support of an overstaffed establishment, operations (such as the cost of power for pumping), and then maintenance.

— Available funds are not necessarily targeted to areas of greatest need, since there is no formal prioritising of work. Regional offices probably allocate funds in proportion to the area served for want of better procedures.

— O&M offers poor rewards and prospects to engineering and technical staff who probably have been trained in design and construction. Staff are therefore often poorly motivated.

— New projects attracting soft international loans have a high profile for all concerned.

No political capital is to be derived from maintaining existing systems. The need for institutional changes in funding, organising and managing maintenance is discussed later.

Maintenance Schemes and Programmes

Despite the fact that engineers have wide experience of system deterioration, in irrigation, as in other sectors such as roads and water supply/sanitation, there appear to be virtually no quantitative data linking the physical and economic performance of irrigation schemes with maintenance programmes. The World Bank estimated that the rates of return to satisfactory maintenance on projects in India could be as high as 40 per cent, but no supporting analyses were presented.

A large number of technical, social, institutional and economic/financial variables may affect the performance of irrigation schemes. It is therefore not a simple task to tie scheme output to a particular variable, such as maintenance. Furthermore, the maintenance needs of different types of schemes are affected by a large number of factors. For example, paddy rice production under supplementary irrigation in humid areas is less sensitive to system condition than upland crops grown in more arid areas, where irrigation provides most of the crop water needs.

In response to the lack of knowledge, Skutsch set out to identify the economic returns to maintenance on two irrigated areas that had recently been rehabilitated under international funding. One project is located in a semi-arid region of South Asia; the other is in the humid tropics. Rehabilitation was undertaken as a response to falling output linked to inadequate maintenance. Detailed information on scheme maintenance spending prior to rehabilitation was not available, but good estimates were derived from government areal maintenance allocations. Crop output norms under medium farm inputs, costs, prices, and farm budgets in 1992 for each area were based on information from the project feasibility reports.

In the initial period after project commissioning, output will be the same under the two maintenance regimes. It is assumed to take five years before the target output of the scheme is realised, as farmers gradually improve their practices. The assumption may be somewhat optimistic, but it does not affect the comparison between the two regimes. Overall, output trends are indicated by smooth transitions. In practice, annual output would fluctuate in response to seasonal conditions (climate, pests, diseases, market prices, etc.) but, again, the simplification will not affect the comparison.

The net benefits of *satisfactory* maintenance over *poor* maintenance are conservatively taken as the value of extra crop output, plus the cost savings from postponing rehabilitation. In fact, further unquantified benefits would include the use elsewhere in the economy of funds needed for rehabilitation; savings in foreign exchange repayments; plus intangible social and environmental benefits resulting from a sustained and more equitable supply.

To avoid the dominating effect on the cost streams of the initial construction cost, differential cost and benefits streams were calculated for each pair of *poor* and *satisfactory* maintenance cases at a discount rate of 10 per cent. The incremental net present value (NPV) attributable to good maintenance, that is the difference between the (benefit – cost) streams for the two maintenance cases, was derived from the production profiles, the unit costs and benefits.

In all cases, *satisfactory* maintenance showed a positive incremental NPV relative to *poor* maintenance, varying from US$470/ha (low value crop) to US$950/ha (high value crop). The equivalent annual incremental net benefits varied from

US$50 to US$100/ha. To set the figures in context, the annual costs of satisfactory maintenance were US$13.5 and US$23/ha in the two cases, for *poor* maintenance the corresponding figures were US$6 and US$9/ha. The savings in expenditure on a medium-sized scheme of 20 000 ha under *satisfactory* maintenance would have been in the range US$1.0-4.0 million, based on discounted per hectare cost advantages of nearly US$50/ha. The capital thus saved on the schemes might have been used, for example, to provide a supply of safe drinking-water to either 7 000 individuals, or 25 000 individuals. Sensitivity analyses showed that the findings were robust when tested against the simultaneous introduction of four discrete and adverse changes, namely:

— discount rate increased to 12 per cent;

— crop prices reduced by 20 per cent;

— maintenance costs increased by 20 per cent; and

— rehabilitation delayed by two years.

No attempt was made to calculate "optimal" levels of annual maintenance because it was considered that there were insufficient data to draw robust conclusions. The chosen discount rate of 10 per cent, as used by the World Bank, appears conservatively high for projects which must be sustained over many years in conditions where local funding is in short supply. In particular, low rates of growth in Africa over the last 30 years, and depressed agricultural prices, which do not truly reflect regional demand, may justify a lower discount rate.

The economic returns to sustained irrigation system performance are also illustrated by an earlier Wallingford study by Chancellor, Lawrence and Atkinson. In this case, a vortex tube extractor was constructed at the head of an existing irrigation system in the Philippines to remove water-borne sediment. Under conventional maintenance procedures, sediment had been removed by dragline, but the annual budget was only sufficient to cover partial clearance of the accumulated material. The vortex tube succeeded in safeguarding supply to the irrigated area, which had decreased progressively over time under inadequate maintenance. The benefits assigned to the structure were taken as the net value of the output from the area that would otherwise have been lost to production.

Despite the use of relatively high discount rates and conservative estimates of expected benefits, both studies demonstrated that there is a high economic return to limited measures, which stem the decline of systems. Poor maintenance causes loss of productive capacity, but also diverts scarce local resources from other sectors of the economy, sets back social progress and worsens existing inequities which, as Howsam and Carter suggest, limits sustainable growth.

Financing and Organising Maintenance

In a changing economic climate, international lenders are increasingly exerting pressure on governments to reduce fiscal deficits. Resources to fund the activities of large, overstaffed irrigation departments have become increasingly scarce. In their turn, governments are applying pressure to their own departments to seek funds from other sources, principally farmers, to help finance operation and maintenance activities.

Farmers see system maintenance as the responsibility of governments. If farmers are required to allocate greater resources to maintenance, either in terms of increased payments or of increased responsibility, they must see a clear incentive in better cash returns, reduced costs or improved control over water. Particularly on rice-growing schemes in monsoon climates, farmers may see little urgency in paying today to avoid increasing problems tomorrow, particularly if the government has traditionally stepped in to rehabilitate defective systems. Realistically, a set of clear and enforceable sanctions against non-contributors and transgressors is also needed. Sagardoy has suggested ways in which governments could establish incentives for good maintenance and penalties for neglect:

— A regular review of individual scheme maintenance standards. Funds for improvements would be allocated only to schemes where maintenance is reasonably carried out.

— Stricter criteria for scheme selection for rehabilitation. If a limiting per hectare budget is established, based on schemes in better condition, those that have been allowed to deteriorate could only undertake work on the most important structures.

— Modernisation projects would only be considered for those schemes with a good maintenance history.

— Farmers should be required to contribute an agreed number of maintenance days per year.

One way of creating an incentive for farmers to contribute their labour for maintenance could be to offer a rebate in water charge fees. A simple system of vouchers earned against workdays could be traded for a discount on fees. Irrigation schemes suffer generic problems resulting from the neglect of maintenance, but the nature and scale of problems vary from scheme to scheme. An Indian Working Group on Major and Medium Irrigation programmes concluded that "... the system of setting O&M rates as a fixed rate per hectare without considering the nature and type of project is not rational...".

It is particularly important to be able to predict the expenditure required to keep a scheme functioning well. If the financial commitments are too heavy in relation to the benefits, farmers will be unwilling or unable to take over new responsibilities, so the scheme may disintegrate. In some areas of the world, small farmers who depend on growing rice for their livelihood may be barely able to survive, let alone increase their contributions to the upkeep of their system, in the face of recent steep falls in international cereal prices. On Indian schemes during the first part of this century, the costs of operation and maintenance, and sometimes also part of the initial capital costs, were fully met from water charges. Gulati, Svendsen and Choudhury related the required water fees in India to gross revenue.

Fees would need to be raised to 5-12 per cent of farmers' income, depending on crop type, to meet the annual costs of O&M, assuming a high rate of collection could be achieved. Another problem was that the recovery ratio, the ratio of water charge receipts to working expenditure, had fallen from over 100 per cent in 1960 to less than 50 per cent by 1990. At the same time, the proportion of the collection spent on administration had increased substantially, from 35 per cent to 48 per cent in the case of one region, over the same period.

The proportion of the total O&M budget contributed by government varies greatly among countries. At one extreme, the Governments of Chile and Mexico provide a small proportion of the total annual O&M cost, while in South Asia national finance pays for between 50 and 70 per cent of the total. The cost of collecting fees may be a significant percentage of the total collection, for example 8 per cent in the Philippines and 15-20 per cent in India.

The World Bank identified five key conditions to be satisfied before fees are increased:

- the managing agency should detail all costs, levels of service and benefits;
- users participate in setting the levels of service;
- users must prove willingness/ability to pay;
- charging mechanism - fees, in-kind contributions or property taxes - to be agreed;
- fees linked to level of service.

The Bank recognised that where local conditions or pricing policies are unfavourable, governments will need to continue to cover part of the costs of O&M. A recognised system of sanctions for non-payment needs to be enforced. A worthwhile improvement in rate of collection requires at least the following conditions:

- an effective information system capable of defining water use in detail;
- a dependable delivery system;

— an agency willing and able to collect fees;

— fees are fed back to improve the system.

System management needs to be transparent to discourage corruption and to promote fairness and equity.

Another method of increasing revenue is to introduce a system of water markets. Rights to water may be traded among users. At the outset, the agency allocates rights and charges holders, who may then sell on the right. In principle, the system not only raises money but also encourages more efficient use of water, as surplus water may be resold by users rather than go to waste. The system requires fairly sophisticated institutional arrangements. Formal procedures in Mexico and Chile, and informal arrangements for trading ground and surface waters in parts of India and Pakistan, appear to work well.

With the backing of international banks and funding agencies, proposals to increase fees are frequently linked to policies for turnover of government systems to farmers. However, there is little information to guide governments contemplating such policies. Turnover involves major changes to existing practices, which will require time, patience, consultation and resources to implement successfully. Turnover may be viewed as a process of moving services closer to the user; more cynically, it may be seen as a way of reducing government expenditure and responsibility. Many questions remain about the design and implementation of suitable structures. No single model will work in widely differing socioeconomic circumstances.

In Mexico, much of the irrigated land has been turned over to farmers' organisations typically farming around 10 000 ha. Turnover is judged a success on the basis that expenditure on maintenance has increased and the condition of infrastructure has improved. There appears to have been no great expansion of the irrigated area, nor increase in unit yields following turnover. Farmers appear to have favoured turnover because it gave them better control over water. The conditions prevailing in Mexico differ from those in many parts of the developing world where turnover is being tried.

In Sri Lanka, attempts to turn over O&M of distributary channels and below, and to increase water charges, were not successful. Farmers had only low returns from small plots of rice and paid little for their water. They saw no incentive to take on extra tasks that had previously been managed by government. The Government was forced to pay them to carry out maintenance at distributary level to avoid the collapse of the systems. In China, "townships" are normally responsible for maintaining distributary channels. Increasingly, irrigation agencies may need to fund works to secure the system against deterioration.

There is some evidence that turnover is more likely to be successful where farmers face some water scarcity, but severe scarcity of supply leads to anarchy, rather than cooperative working. There is a danger that they will adopt a short-term view, to the detriment of the system. Turnover, as envisaged in many countries, leaves the government responsible for expenditures on much of the main system, representing a large component of the total O&M cost and the greater part of the problem. Farmers have little reason to make turnover work unless they get an improved service from the main system. At present there appears to be no concerted effort to improve the way maintenance is undertaken in the main system. Bank-supported policies to increase water fees and to turn over systems in these circumstances are potentially setting at risk huge investments in irrigation infrastructure.

Private Firm or Autonomous Agency

A third party, either a private company or an autonomous government agency, could manage operation and maintenance of the main system. Under financial autonomy, the irrigation agency or company would be responsible for obtaining resources and controlling their use.

It is unlikely that established private companies would be interested in the returns to be made from rice cultivation, although there may be potential with higher value crops. Attention needs to be given to the potential offered by autonomous maintenance agencies. The experience of the

World Bank with rural road maintenance organisations suggests that decentralisation can be successful. Success depends on providing appropriate incentives, acquiring and keeping the right staff.

It is likely that government would need to provide some subsidy to make an autonomous maintenance organisation viable on rice-growing schemes. Realistic assessments of the fee for water would be needed. The organisation's income, including any subsidy, would need to allow a fair profit, provided that a high rate of collection was achieved to allow for the costs of regulation. Formal contracts between farmers and the agency would allow a measure of control, since the agency would derive a large part of its income from users. However, the organisation would enjoy a monopoly position so would need regulation against performance targets. For example, it would be required to irrigate the full feasible service area and to reinvest in the infrastructure. Otherwise, there would be a temptation to maximise revenue by, for example, reducing investment and providing a reliable supply to a limited area at times of shortage, rather than trying to spread water equitably across the command.

Assistance from Lending Agencies

The reluctance of donor agencies to fund recurrent costs stems from concerns about accountability and a recognition that support for O&M would merely delay the time when nations would have to assume responsibility for the expenditures. It has been suggested, that lending arrangements might include a transitional period after construction during which initial support for O&M was provided, but progressively decreased over time. The loan would be conditional on the establishment of appropriate domestic institutional arrangements to manage O&M. Experience of such arrangements, for example in Pakistan, does not seem to be positive. It has been suggested that donors might link new funding to the performance of existing projects, focusing greater attention on project monitoring and *ex-post* analysis. It is likely that such ties would be unacceptable to some recipients, and banks might also have difficulty in introducing corresponding uncertainties into their lending programmes. Lending agencies have generally turned away from transitional support in favour of turnover policies.

Improving the Effectiveness of Maintenance

Maintaining irrigation systems at a *satisfactory* level is consistently more economic, for both high and low value cropping patterns, than allowing systems to deteriorate under *poor* maintenance. If this *common sense* approach is applied at the outset then money could be more productively employed in other areas of the economy rather than prematurely rehabilitating the system. The extra spending on a medium-sized project of, say, 20 000 ha might otherwise have been used to provide a safe supply of drinking-water to a small town. To improve the effectiveness of maintenance, irrigation planners, managers and their political masters must recognise the impacts of neglect, adjust policies, and think about incentives, institutional changes, design and planning procedures.

Policy

Governments are under pressure to reduce spending. Fees for water generally need to be raised and the rate of collection improved. Such changes are difficult to introduce without organisational change. Guidance in restructuring inefficient institutions to manage maintenance better is needed as there is little informed advice on the consequences of different policies. Turnover, as a strategy to reduce government spending, is being pursued without a proper understanding of what the impact of maintenance by farmers will be. Larger systems have been designed on the assumption that an agency employing trained staff would be responsible for O&M to the level of the tertiary offtake, and that the same body would have some measure of control/ influence over O&M at the tertiary level. Under turnover, that will no longer be the case. Different groups of farmers will undertake O&M as they see fit, in different parts of the system. Guidelines for governments based on case studies, practical training for farmers and simple manuals are needed to ensure that the infrastructure is maintained, and to identify the support that farmers will need.

Incentives and Penalties

Innovative ways of introducing incentives to carry out maintenance, together with clear penalties for failure, are needed. Schemes mostly remain under the management of irrigation staff who may be frequently transferred, and therefore have limited interest in future improvements. Farmers should not be penalised for the poor performance of the agency. One way of encouraging maintenance of the lower levels of the system could be to offer a rebate in water fees against the number of days a farmer works on maintenance. A simple system of vouchers for the number of days worked could be traded for a discount on fees.

Institutional Change

Partial turnover to farmers will not address problems of maintenance on the main canals of most large systems. It is crucial to undertake organisational changes at that level of the system as farmers are unlikely to cooperate unless they are convinced they will get an improved service. In many parts of the world it is increasingly difficult to retain experienced field staff on government-run irrigation schemes. Cadres of experienced staff and workers are reaching retirement age. The outlook for sustaining the infrastructure in a workable condition appears serious. Unless incentives and training for staff and workers can be improved, the situation can only get worse.

The potential for decentralising maintenance to autonomous, self-accounting units needs wider consideration. An organisation dependent on water charges for much of its budget has a strong incentive to improve collection rates. Some government subsidy may be required on rice-growing schemes to make enterprises financially attractive, and units will need to be regulated to ensure that they are investing in the whole system. The government contribution would be dependent on *satisfactory* performance by the unit. Experience from a number of countries suggests that effective decentralisation of a huge bureaucracy may require a period of decades. Changes are needed now, so franchising arrangements might be necessary.

Design

Designs for low maintenance giving easier and less costly management, reduced maintenance, lower service fees and greater effective life need more importance during project preparation. In the future, systems will be operated and maintained by farmers. Engineers and planners need robust information, comparable to that now available in the roads sector, on the links between design, maintenance spending, performance, whole life cost and sustainability. Field studies in a number of countries/regions are needed to provide facts on the deterioration of irrigation systems and maintenance performance, so as to get rational policies.

Planning Procedures

Better identification of maintenance works and priorities are required to make the best use of limited funds. In the face of growing water shortages and increased competition for available water with other parts of the economy, managers will need to introduce improved practices and planning methods, drawing on asset management procedures being widely introduced in other parts of the water sector.

References

Finney C E., "The agricultural benefits of sub-surface drainage on irrigation schemes", *ICID Journal*, New Delhi, India, 1997.

Johnson S H., *Irrigation management transfer: decentralising public irrigation*, IIMI, Mexico, 1993.

Skutsch J C., "Maintaining the value of irrigation and drainage projects", *Report OD/TN/90*, HR Wallingford, Wallingford, UK, 1998.

13

Management of Sprinkler Irrigation System

Sprinklers produce more yield than typical surface irrigation systems per ha-m because sprinklers apply water more efficiently. Not all water applied by an irrigation system is used by the crop. Some water is lost to deep percolation, evaporation, or runoff. Application efficiency (Ea) is a term that tells how much of the water applied by the system is actually stored in the root zone for crop use.

A good sprinkler system has an Ea of about 70 percent which means that 70 percent of the water applied by the sprinkler heads is actually stored in the soil for crop use. The actual Ea depends upon how evenly the sprinklers distribute water and other factors such as operating pressure, nozzle size and spacing, wind, air temperature and humidity (day versus night), irrigation scheduling and maintenance condition. The average efficiency of surface irrigation in the intermountain West is probably less than 50 percent as compared to the higher sprinkler efficiency of 70 percent.

Efficient Irrigation System

An efficient sprinkler system is the result of good system design, proper irrigation scheduling, and careful operation and timely maintenance.

Design

A well designed sprinkler system applies water uniformly to the soil surface and is capable of applying enough water to meet the peak demands of the crop without producing excess runoff. Good design considers such factors as pressure; nozzle size and spacing; wind, air temperature and humidity (day versus night); soil intake rate; crop rooting depth and water use rates.

The flow rate from a sprinkler nozzle depends upon nozzle size and water pressure. Flow rates for selected nozzle sizes and pressures are given in Table 1. Typical sprinkler flow rates may vary from four gallons per minute (gpm) from a 5/32-inch nozzle at 30 pounds pressure to over 11 gpm from a 7/32-inch nozzle at 70 pounds pressure. The nozzle size is usually stamped on the nozzle. Wheelmove systems typically have 3/16-inch nozzles.

Table 1. Sprinkler Pressure and Flowrate

Nozzle size	*Nozzle Pressure, psi*				
	30	*40*	*50*	*60*	*70*
	inch Nozzle flow rate, gallons per minute (gpm)				
5/32	3.9	4.5	5.0	5.4	5.8
11/64	4.7	5.4	6.0	6.6	7.1
3/16	5.5	6.3	7.0	7.7	8.3
13/64	6.4	7.4	8.2	9.0	9.7
7/32	7.4	8.6	9.6	10.5	11.3

On sloping fields there may be considerable pressure differences between sprinkler heads on high and low ends of the line. In this situation, flow control nozzles may be used to improve the on sloping fields there may be considerable pressure differences between sprinkler heads on high and low ends of the line. Flow control nozzles apply water at the same rate, regardless of water pressure.

Precipitation Rate (Pr)

The Precipitation Rate (Pr) is the rate at which water is delivered from the nozzle, averaged as inches per hour, over the area covered by one nozzle. It is important to consider the Precipitation Rate when designing a sprinkler system, since water will run off if applied faster than the soil can absorb it. Precipitation rate can be calculated using the following formula:

$$\text{Pr (inches/hr)} = 96.3 \times \text{nozzle flow rate (gpm)/area covered (ft}^2\text{)} \quad (1)$$

Precipitation Rate can be calculated as follows: In a typical wheelmove system, each sprinkler covers 2400 square feet. This is based on a spacing of 40 feet between sprinklers on the line, and a 60 foot move (40′ x 60′ = 2400 square feet). With 3/16 inch nozzles that are operating at 50 pounds pressure, the nozzle flow rate is 7.0 gpm (from Table 1). The Precipitation rate would be:

$$\text{Pr} = 96.3 \text{ (7.0 gpm)/2400 ft2} = 0.28 \text{ inches per hour} \quad (2)$$

Application Rate (Ar):

The Application Rate (Ar) is the average rate at which water is stored in the soil, in inches per hour.

$$\text{Ar} = \text{Application Efficiency (Ea)} \times \text{Precipitation rate (Pr) size} \quad (3)$$

While sprinkler application efficiency varies with field conditions, 70 percent is a reasonable value.

Example

$$\text{Ar} = (70/100) \times 0.28 - 0.20 \text{ inches per hour}$$

How Long to Irrigate (Duration)

The duration of irrigation needed to store the crop irrigation requirements in the root zone is:

Irrigation Duration (hours) = Crop Irrigation requirements (inches)/Ar (4)

Example

Determine how many hours to irrigate in July with a crop irrigation requirement of 8.5 inches, 3/16 inch diameter nozzles at 50 psi and 40′ x 60′ spacing (use results of previous examples).

Hours to irrigate in July = 8.5 inches/ 0.20 inches/hour – 43 hours

Assuming that the sprinklers were moved twice per day, about 11 ½ hour sets, then about four irrigations (4 – 43/11.5) are needed in July. This is equivalent to one 11 ½ hour irrigation about every 8 days.

Calculated irrigation duration for nozzle sizes of 5/32 to 7/32 and pressures of 50 and 60 psi. The use of Table 1 and Equations 1, 2, and 3, assuming sprinkler spacing of 40′ by 60′ and 70 percent application efficiency.

Operation and Maintenance

To realise the full benefit of the sprinkler system, it must be operated according to design and properly maintained throughout the irrigation season. This may involve special operating techniques such as using an offset hose or alternating between day and night on successive irrigation cycles to improve distribution uniformity. Where pressure differences within a sprinkler system result in low uniformity of water application, special hardware such as flow control nozzles or pressure regulators may be required.

Typical hand move and wheelmove sprinklers use impact sprinklers with 3/16″ diameter nozzles operating at 45 to 50 psi pressure. Unfortunately, some systems are operating at pressures somewhat below this which lead to problems with distribution uniformity.

Several years back, in a program funded in part by the Utah Department of Agriculture and the Utah Agricultural Experiment Station, numerous sprinkler irrigation systems around the state were evaluated. Irrigation application efficiencies (water stored in the root zone divided by a water delivered to the field) for wheel moves varied from 39 percent to 75 percent. With reasonable good management, the wheelmove application efficiencies averaged nearly 70 percent.

The lower application efficiencies were from lower pressures than designed or, perhaps, a greater distance in the move than is desirable. One such system we evaluated had a coefficient of uniformity (CU) of 62 percent on the 40 foot by 60 foot sprinkler spacing. The uniformity improved to 87 percent when a 20 foot offset was used on alternate irrigations. The 20 foot offset, in this case, was equivalent to one revolution of the wheel move to either the right or the left of the previous irrigation lateral position.

An audit or evaluation of the irrigation system is recommended if you suspect that the system is not as efficient as it might be. An audit determines application depth, distribution uniformity, and hydraulic performance of the supply system. If a pump is used, it is tested to determine fuel or energy use efficiency. An audit may also identify steps to improve system operation and maintenance.

Good operation also includes matching the set time (or rotation time with a center pivot) with the applied water depth and application rate to maximise the fraction of water stored in the root zone. Field irrigation (application) efficiency is the water stored in the root zone divided by the water delivered to the field. If a field is under-irrigated, a high irrigation efficiency could result with a low uniformity. Conversely, an over-irrigated field will have a low irrigation efficiency even with a high uniformity because of the deep percolation. Thus, a knowledge of the soil moisture content prior to irrigation is essential to maintaining a high application efficiency while providing for optimum crop water use and growth.

Wheelmove Sequencing

The objective of the moving sequence is to maintain adequate soil water for best crop growth, while not causing excess deep percolation. Deep percolation is detrimental to water conservation and salinity reduction goals. Various wheel move sequencing scenarios have been tried and/or suggested:

a) Irrigate every move across the block and then roll it back empty. This move sequence is similar to the way hand move lines were used. But a major disadvantage is the long distance the empty line is rolled back.

b) Skip irrigate every other move across the block and then on return irrigate the skipped sets on the way back instead of rolling back empty.

c) Irrigate every set across and then start, say twelve hours later, and irrigate every set coming back.

   This avoids rolling the line empty all the way back, it also avoids some of the concern about the skip irrigation having alternating unirrigated spots on either side that may bother people. However, irrigating so soon after the previous irrigation causes extra deep percolation losses on the wet side of the field. There is also a longer than needed interval between irrigations on the opposite (or "dry") side. This mode is not recommended due to the alternate over irrigation followed by soil water shortage

D) Irrigate every set across, roll the line empty for about three sets back and then irrigate the rest of the way back. Roll the line empty for three sets or so and then irrigate across again.

e) Other possible variations exist and depend upon soil, crop and field characteristics.

Wheelmove Maintenance

Much of the maintenance is related to the single cylinder air-cooled engine and associated hydraulics, gears, and chain drive for moving the system. It is important to follow the engine manufacturers' recommendations for oil viscosity and change intervals and for air filter, spark plug, and other service requirements.

The wheelmove manufacturers provide adequate operation and maintenance instructions to fit their particular drive system. It is important that the owner/operator read and be familiar with and follow the instructions that match the equipment he/she is using.

Check all nozzles and impact sprinklers for plugging, mismatched sizes, breakage, corrosion or other damage caused by wear or winter weather. Couplers and connections should be checked for leaks and repairs/replacements should be done early. It is a good practice to identify problem components at the end of the previous irrigation season and to have the replacement parts on hand for spring installation. If water leaks occur at joints or drain plugs during irrigation, check the gaskets and pipeline connections for wear or cracks and replace them as needed. Check and tighten the couplers and connectors as required.

In those places where an open ditch water supply is used, adequate screening and sediment removal are essential for minimum trouble operation . This is particularly a problem in the spring and early fall with more trash in the ditches and canals at those times.

There are quite a few anecdotal stories about runaway wheelmoves in high winds. Anytime the wheelmove is empty, and particularly during the off season, it is essential that it be anchored to prevent wind damage. Often this is as simple as moving the system over to a fence and tying it down in three or four places to sturdy posts. However, special wind anchors are available which attach to the lateral pipe and act like a brace to prevent movement.

Gravity Pressure Sprinklers

Wheelmoves and hand moves are used on almost all of the more than 300 gravity pressurised sprinkler systems in Utah. Utah's mountain valley topography is favorable to developing gravity pressurised pipelines. Much of our irrigated area is in mountain valleys in close proximity to canyon streams. Thus, it is often economic to install a pipeline up the canyon to gain about 110 feet or so of elevation induced head. This also has the advantage of reducing open channel water seepage losses in the canyon mouth alluvium and extends the available water supply.

Many previously surface irrigated systems (private irrigation companies as well as individuals) have converted over to gravity pressure wheelmove irrigation. This switch from the traditional (often pioneer built) surface irrigation to gravity pressurised sprinkler has created some interesting situations. The value of "head end" water access opportunities in preference to a "tail end" location in multiple user surface irrigation systems is well known.

However, when a conversion to gravity pressure sprinklers occurs, the historically "tail end" irrigators become switched to the "head end" position in that they now have the highest pressure. For some, this is the first time ever that they have had access to abundant water instead of the tail end dribble they are used to. These advantages are often short-lived as the irrigation company board of directors have responded by requiring pressure or flow regulated nozzles and strict adherence to water deliveries proportionate to water stock shares owned.

Soil Moisture Management

Agricultural researchers in the Western U.S. have studied the effect of soil water availability on crop yield through the use of such techniques as sprinkler line source field research plots. In this research the irrigation water is varied from none to excessive amounts.

Irrigation Scheduling

Irrigation scheduling is the process of determining when to irrigate and how much water to apply. It depends upon design, maintenance, and operation of the irrigation system and the availability of water. The objective of irrigation scheduling is to apply only the water that the crop needs, taking into account seepage, runoff losses, and leaching requirements. Scheduling is especially important to pump irrigators if power costs are high. Common irrigation scheduling approaches include the following:

1. Irrigation on fixed intervals or following a simple calendar, i.e., when a water turn occurs or according to a predetermined schedule.
2. Irrigating when the neighbor irrigates.
3. Observation of visual plant stress indicators.
4. Measuring (or estimating) soil water by use of instruments or sampling techniques such as probes.
5. By following a soil-water budget based on weather data and/or pan evaporation.
6. Some combination of the above.

For irrigation scheduling to be most useful at a specific location, the following should be done:

1. Evaluate the irrigation system. Determine application depth, efficiency, and operating capabilities and constraints.
2. Select an appropriate irrigation scheduling method.
3. Monitor performance at intervals during the growing season.
4. Perform a post-season evaluation and determine changes for next year.

As a general rule, field crops should be irrigated whenever the soil water depletion approaches 50 percent of the available water in the root zone. In the peak crop water use period in an arid area, the occurrence of rain is often neglected in determining an irrigation schedule. This field did not receive sufficient irrigation from the third week of June through late July and again in late August. The soil water content was below 50 percent of available soil moisture prior to and following the first cutting on July 6 and prior to the second cutting on August 30. Consequently, the crop was water stressed and the yield was reduced about ¾ ton/acre below what could have been realised with better irrigation scheduling.

References

Bauder, J.W., *Simple Yield Models Based on Water*, Montana State University, Bozeman, Personal communication, 1998.

Hill, R.W. and R.G. Allen, "Simple Irrigation Scheduling Calendars", *Journal Irrigation and Drainage Division*, ASCE. 122:2, March/April, 1996.

14

Organisation of Irrigation Schemes

A segregated organisational structure is made up of a number of units acting independently, but in a coordinated manner, to achieve the final goal of the betterment of the farmers' welfare. In other words, there is one organisation that provides for the appropriate use of water, another, or several, for the appropriate use of other inputs, and others for the remunerative selling of the produce, etc. Although this appears a highly desirable approach since it allows the development of each organisation according to the specific needs, experience shows that all too frequently the establishment of the organisations related to inputs other than water never materialises. The reason for this is that water is needed by every farmer and the corresponding organisation is established in one way or another as soon as water becomes available; however, the need for the other organisations (marketing, cooperatives, etc.) is felt in so many different ways by the individual farmers that the sheer diversity results in the fact that they are not established. This, indeed, handicaps the achievement of the final goal. In order to get a segregated organisational structure working properly, it is necessary to motivate and promote the formation of the organisations that provide the other needed supporting services, and this often has to be done by the government.

The characteristics of the organisations needed to achieve an appropriate use of water will be described later in the text.

Other organisations, such as cooperatives for different purposes: marketing, machinery, production, savings, social organisations, etc.,

INTEGRATED ORGANISATIONAL STRUCTURES

Integrated organisational structures are those where the final goal is achieved through the interaction of several departments or organisational units, each one aiming at lower goals, all connected through a clear line of command. These organisational structures are frequently referred to as "project authorities" or "project organisations", since they deal with most of the socio-economic issues of an irrigation project in an integral manner.

An organisation aiming at covering all the many aspects of irrigated crop production, marketing and socio-economic factors must have a very complex structure to cover all the different situations. Therefore the first, and perhaps, most important, condition for the establishment of these organisations is a certain degree of homogeneity in the socio-economic conditions of the farmers. This is the main reason why this type of organisation is mostly found in settlement projects, state farms, state cooperatives, and suchlike. Otherwise the complexity of the administration is such that the chances for failure are too high to be taken.

Integrated organisational structures are useful to overcome difficulties in coordination at national level. For instance, it is frequently mentioned in project evaluations that the credit system through normal banking' institutions only reaches a minority of the irrigators, due to lack of sufficient coordination between the banking institutions and the irrigation projects. To remedy this, in some settlement projects, the organisation provides the farmer with all the necessary inputs on a credit basis and at the end of the harvest all the credited inputs are subtracted from the total income and the rest returned to the farmers. Such measures are indeed a guarantee that all irrigators enjoy equal opportunities, as far as credit is concerned. The same can be said about other necessary services such as extension, applied research, etc. However, because the establishment of integrated organisational structures may duplicate the functions undertaken by existing institutions, special care should be taken to analyse the potential risks of overlapping of any functions and activities.

Integrated organisational structures may be a good way to solve problems of coordination at the national level, but in order to be able to do so they must be well-organised and equipped with competent technical staff and sufficient financial means to undertake their many responsibilities successfully. This implies firstly that there must be a central/national institution capable of procuring the technical and financial support necessary at the project level and, secondly, that the government must be willing to delegate a considerable amount of authority and autonomy to the individual project authorities. Otherwise there will be permanent conflict between the project authority and the departments of the national institutions dealing with credit, research, extension, etc.

Types of Irrigation Organisations

The criteria used are only one way of categorising most of the existing irrigation organisations, while other authors use different criteria. The main criterion chosen to classify the organisations is whether the organisational structure covers all the development activities (water management, agricultural extension, applied research, supply of inputs, credit, marketing and basic infrastructure and social services), or a few of these activities, or only those related to water management. Within each of these categories a further distinction is made depending on the degree to which management is controlled by the farmers and/or the government. The followng are the main types of irrigation organisations:

- Integrated management organisations
- Specialised water management organisations

— Multipurpose water management organisations.

No attempt is made to cover all the organisational possibilities but to illustrate the salient features of the most characteristic ones. For greater details on the described organisations or information on other organisational possibilities, the references should be consulted.

Integrated Management Organisations

These organisations are characterised by an integrated organisational structure where all development activities are undertaken by specialised units which are all connected by a clear line of command and finally responsible to a single person (project manager) or committee. There are several types of integrated management organisations depending on the degree of intervention of the government and the farmers in the management of the scheme. Where the organisation is run by the farmer it usually adopts the shape of a Service Cooperative. Where the government officials fully control the organisation there are two possibilities depending on the size of the production units:

— State Farms (large production units)

— Irrigation Settlement Projects (small production units)

Another distinctive feature of these cooperatives is that most of the farmers involved are relatively wealthy. This is almost a precondition because of the high cost of irrigation infrastructure. Although part of the needed funds can be obtained through loans, it is obvious that such credit must be backed by the strong payment capacity of the farmers.

The organisation of an irrigation cooperative depends greatly on the kind of contributions made and the services needed by the members. The organisational structure is similar to that of the Irrigation Association which is described later in the text. The rules and regulations governing the activities of the cooperative are established within the framework of the national laws for cooperatives.

A very distinctive feature of a cooperative is that it is established by the free will of its members and is only binding on them, while many irrigation organisations are established by a majority of the members and can be binding on the minority, who sometimes may not be members of the association.

State Farms

State Farms are often established where the land is nationalised or where land reform processes have taken place. The purpose of establishing a State Farm can be to maximise agricultural production or to gain experience on newly reclaimed lands for its later transfer to farmers as settlement projects. The first type is the most commonly found.

Much disappointment exists all over the world with the performance of State Farms. Since all the production inputs can be provided at their optimum level one would expect that State Farms would be able to achieve the highest production levels. However, it is often found that the production levels are higher in private farms and their operating costs are similar to those of the State Farms. There are even cases where State Farms operate at higher cost than settlement schemes.

The management of a large State Farm, with all the numerous and complex activities that it involves, is extremely difficult under the limitations usually imposed on public institutions. Late release of government funds for operating the farm, unavailability of inputs at the time needed, lack of involvement by the government officials in the potential benefits, and large bureaucracies, are some of the many reasons which often hamper the management of State Farms.

In order to maximise the productivity, there is a noticeable tendency to use highly capital intensive technology for the water distribution systems and machinery for the agricultural practices. This in turn means that the management of these farms must be highly technically qualified. Unfortunately such personnel are not abundant in developing countries and they are more attracted by positions in the main offices than by those in the field. Insufficient properly trained personnel is often a serious limitation for the development and operation of State Farms.

The organisational structure of State Farms is characterised by a number of units covering the main development and production activities and directly dependent on the Manager or Executive Director. The question of whether the farms should be subdivided into several independent production units or grouped into a large central unit is often discussed. A convenient solution is to provide most of the services under a central unit with the exception of those related to the main agricultural inputs (water, fertiliser and machinery) which are normally integrated in farm production units.

The manpower and farm machinery requirements of the State Farms are determined by criteria of maximum efficiency. However, in practice the previously mentioned management problems lower the performance standards to values often below those encountered in less centralised organisations.

Irrigation Settlement Projects

The most complex of the Integrated Management Organisations are the irrigation settlement projects, described here in more detail. Irrigation settlement projects aim at improving the economic and social welfare of landless people or poor farmers by providing them with irrigated land and agricultural production means.

The greatest difficulty in managing an irrigation settlement project successfully arises from the fact that most of the settlers have a low educational level, practically no financial resources and, occasionally, even no experience in agriculture. To upgrade them from their subsistence level to farmers living on commercial agriculture is a huge task, and one should not be surprised at the effort and time needed to achieve this.

The key to the success of a settlement project is that government officials must initially be responsible for and perform many of the tasks, only giving the farmers simple and straightforward tasks which they can carry out with their own means. As time goes by, it may be possible to transfer more and more of the tasks and responsibility to the farmer.

The question arises as to how to integrate or coordinate such a wide variety of services needed for agricultural production, and often under different ministries. The formation of a coordinating committee in charge of bringing together the responsibilities of the different institutions is, in theory, the easiest way. However, such an arrangement seldom works in developed countries and much less in developing countries where financial and organisational problems are greater. Therefore, the solution of a single integrated organisation providing most of the social and technical services is normally preferred.

The choice of the alternative depends on the existence or absence of a national institution specifically designed to support and financially back-up the integrated organisations at the project level. Where such national institutions exist, an integrated structure generally functions better than a segregated one. In this connection, it should be noted that an integrated organisation is most needed during the initial stages of a settlement project. Once settlers become more independent, because of the higher standards, many management responsibilities can be transferred to them or cancelled. This learning period lasts for many years—frequently 15 or 20 years—but can hardly be avoided.

For obvious reasons of complexity of the tasks to be undertaken and financial requirements these projects are managed by government officials and therefore can be considered as public institutions during the initial stage.

i) *The structure:* The organisational structure of settlement projects generally has the following main units:

a) *The Manager's Office*: This is the highest executive body of the scheme and frequently consists of the Manager himself and some advisory members. He is directly dependent on the national institution responsible for the management of all irrigation schemes in the country.

b) The executive units, which are generally:

— operation - finance

— maintenance - marketing

— training - production

— administration - social assistance

ii) *The development of a settlement project by stages:* Settlement projects generally go through several identifiable phases, where the project organisation has to adjust to the main tasks to be performed: (a) planning; (b) implementation; (c) transitional stage; and (d) full development.

However, the above phases only apply to settlement projects where the farmer may become the owner of the land at the end of the process. For those settlement projects where farmers are permanent tenants, there is little difference between phases (c) and (d).

a) The planning stage of a settlement project does not differ much from that of other irrigation projects, with the exception that some additional problems have to be considered and studied, such as (1) size and type of farm; (2) settlement villages with all their infrastructure; (3) communications; (4) workshop for maintenance of equipment; (5) stores; (6) regulations for the future land tenure; (7) organisation. Most of them are treated in depth in the specialised literature. No particular organisation is needed at this stage, since the planning of these projects is normally done by the government institutions and/or consulting firms.

b) The implementation stage is the period between the beginning of the construction work and its end. During this period the organisation of the project is rather variable. Activities normally start with some of the implementation tasks, such as land development works, construction of on-farm structures, establishment of experimental plots, etc., moving gradually towards those tasks more intimately related to the management of the scheme.

This procedure is found appropriate since it allows the project authority to gain knowledge about the physical characteristics of the area and to anticipate some of the problems that may arise later.

Of particular importance during this stage is the selection of settlers, since the future success will depend largely on the farmers selected. The importance and magnitude of the services needed will also depend on the farmers selected. To evaluate their capabilities questionnaires can be utilised. In addition to an initial careful scrutiny, settlers can be subjected to probationary periods of up to 5 years. This practice has been widely adopted in Spain and has been found to be useful. It not only allows the management to evaluate the farmer thoroughly, but also the farmer himself may realise his capabilities of becoming a small "entrepreneur".

Settlement projects are frequently established in newly developed areas where uncertainty exists about their potential and the appropriate techniques. In such circumstances, a state farm is often established during this stage of implementation. After some years of operation, when the land potential has been fully developed, the state farm is converted into a settlement project. In this case, to avoid losing the experience of the staff working on the state farm, part of that staff should be retained in the management of the settlement project.

c) The transitional stage is the period between the completion of the public works and the full development stage. It often involves a long period of farmer training, but the nature of the training varies in accordance with the government's ultimate objectives with regard to the form of the project at the fully implemented stage. In certain projects, particularly those set up for the specific purpose of producing high-value cash or export crops, the government may aim to maintain an "integrated management system" for a long period. In that case, the emphasis of farmer training may be largely on improved agricultural techniques. Elsewhere the objective is likely to involve the devolution of increasing responsibility and freedom of choice to the farmers; and where the government's intention is to set up IAs, the transitional stage will be as much concerned with training farmers to organise themselves to undertake the major water management functions as with training them in agricultural practices.

In integrated management schemes, government retains control over a wide range of activities. In the Mwea scheme in Kenya, for example, the official project management has control over all the following areas, besides the initial selection of farmers and the enforcement of irrigation rules:

- land tenure (land is owned by the State; farmers' tenancies are renewable annually and subject to termination if their performance is deemed unsatisfactory);
- choice of crop (monocrop rice, which is sold through a single marketing channel controlled by the project management);
- timing of cultivation operations (official management organises mechanised land preparation according to pre-planned schedules; irrigations to each field are organised by project staff, not by farmers).

In the first 10-15 years of its operation the Mwea scheme, which is an extreme example of a highly supervised project, has been very successful in terms of increased production and higher farmer incomes. However, experience elsewhere has tended to show that on projects where the mechanisms for centralised control are particularly powerful in the initial stages, there is a danger that their institutions fail to adapt to circumstances changing during a period of time, with consequent social and economic stagnation. Insecurity of land tenure becomes a particularly inhibiting factor, since farmers have no long-term interest in investing in their own land. A system common on many post land-reform projects, whereby farmers are allowed to purchase their land over a 20 or 30-year period, provides for a similar degree of management control in the initial stages but has the great advantage of allowing farmers to assume increasing responsibility for decisions at the farm and local group levels.

Such a policy is followed in Spain, where new settlers are probationary tenants during the first 5 years of the transitional stage but are subsequently allowed access to a permanent title to their land if they satisfy the government authorities with regard to their standards of farming and irrigation. The period of preparing farmers for the full development phase, when they are supposed to take over the management of the scheme as an IA, can often take between 10 and 20 years.

d) The full development stage is the stage at which the potential of the area has been fully developed in technical as well as socio-economic terms. Where the purpose is to establish independent IAs, a gradual transfer of responsibility can be achieved either by geographical sectors—those more advanced and better organised will start and be followed over a period of time by subsequent ones—or by delegating functions over a period of time. Usually the geographical transfer gives better results.

Where government wishes to retain overall control of the management of the scheme but is not concerned with operating an integrated management system, the original organisational structure of the scheme will have to be modified. Many of the activities that were initially important may no longer need to be carried out by the project itself and can be left to other organisations, e.g. research, credit and possibly extension.

Irrigation Cooperatives

Many types of cooperatives exist but here reference is made to a particular type in which a group of farmers associate to develop a common irrigation system for their properties, and jointly farm the land. Such a type of cooperative occurs mostly among progressive farmers and is generally found in developed countries. It is often financially attractive to farmers because they enjoy the advantages (tax exemptions or reductions) applicable to any other cooperative.

Specialised Water Management Organisations

A specialised water management organisation can be broadly defined as a social organisation aiming at an appropriate use of water for irrigation purposes among the farmers of a community. The term "appropriate" is used here to designate a timely and equitable distribution of the water.

The functions of specialised water management organisations are: the operation of the irrigation and drainage systems, the maintenance of the systems, and the assessment and collection of water charges. They are supported by certain general management services (finance, personnel, planning and monitoring).

Mention should also be made of the fact that some of the water management organisations are entitled to the use of water for purposes other than agriculture. This is the case in the USA irrigation districts, and those of Hungary. These purposes may include any of the following: hydropower generation, navigation, water supplies for towns, villages and industries. The question whether or not an irrigation scheme should limit its functions to the agricultural uses of water is somewhat controversial but, provided that irrigation remains the main use, there is no harm in undertaking some of the above mentioned functions. On the one hand, this helps the government administration to simplify its bureaucracy and, on the other, benefits the irrigation scheme with some extra income. Of course, such a system can only be established in conformity with the existing laws and institutions.

The degree of intervention by the government and the farmers in undertaking the functions described above indicates the main types of specialised water management organisations, which are:

i) Organisations mainly controlled by the farmers: There are many relevant examples of this type: the Irrigation Districts (USA), the Water Districts (USA), the Comunidades de Regantes (Spain), the Subak Irrigation Association (Indonesia), the Irrigation Associations of Afghanistan, the Asociaciones de Canalistas (Chile), etc. They are termed here as Associations of Irrigation Water Users or, more briefly, Irrigation Associations (IA).

ii) Organisations mainly controlled by government officials: These are also very numerous and are found in many countries (Spain, Turkey, Bolivia, Iraq, most of Ecuador and Kenya). They are denominated here as Public Irrigation Schemes (PIS).

iii) Organisations with mixed control by farmers and government officials: Schemes where the main irrigation system is managed by government officials while the tertiary canals are controlled by farmers' associations, are increasing in several Asian countries. They are designated as "Irrigation Schemes with Mixed Control".

Although countries tend to adopt policies favouring either (i) or (ii), it is not infrequent to find both types in the same country.

Irrigation Associations

The most distinctive characteristic of Irrigation Associations (IA) is that they are organisations of the people, for the benefit of the people. However, even though government participation is excluded by their very nature, most of these associations would never have existed without the promotion and encouragement of governments.

i) *The structure:* From the organisational point of view an IA is made up of the following executive bodies:

   a) The General Assembly composed of all the farmers of the association. It is the highest authority. Its main function is to select their representatives (Board of Directors) and to approve or disapprove the management plans.

   b) The Board of Directors: This is the highest executive body. Its main function is to supervise and direct the execution of the work approved by the assembly and to prepare management plans annually.

   c) The Manager's Office: Is directly responsible for the execution of the mandate giver by the Board of Directors and for day-to-day work.

   d) The Executing Units: Are responsible for specific functions, such as operation, maintenance and administration.

   e) The Irrigation Juries: Some of these associations—especially the traditional ones—have their own juries to punish faults against the set rules and regulations. The jury is generally selected from distinguished members of the Board.

ii) *Coordination with the supporting agricultural services:* The very existence of a water management organisation implies that a segregated organisational structure has been selected and that therefore the other necessary services are provided by other institutions. All too frequently this assumption is arrived at too quickly since the mere presence of extension, credit and other services in the project area does not mean that they are available to farmers in sufficient quantity to satisfy their specific needs. This is particularly important in the case of irrigation extension services which are a prerequisite to proper use of water and other agricultural inputs.

To ensure coordination with the agricultural services and to engage them actively in the affairs of the irrigation scheme, it is good practice to include representatives of such institutions on the Board of Directors of the association as special members. Alternatively, representatives of the IAs should be invited to the decision-making meetings of the concerned institutions at regional level or, where justified, at national level. In any case, it is important to assure the possibility of a dialogue between the institutions providing the services and the representatives of the farmers receiving them.

iii) *Main characteristics*: This includes their origins, main types (modern and traditional), legal character, water use rights of the association and its members, operation of their main functional bodies and financial responsibilities. Reference is made here only to their most outstanding characteristics:

a) Irrigation Associations (IAs) secure farmers' participation in decision-making for the irrigation schemes. The democratic process of selecting farmers' representatives guarantees such participation. In a world where the people's desire for participation in public affairs is expanding quickly, securing such participation is most relevant.

b) In this connection it may be opportune to raise the issue of the non-political nature of these associations. Their development has sometimes been handicapped because of the belief that these associations may become political and so be a powerful instrument of political leaders. Although in theory such a risk exists, practice has demonstrated that it rarely happens.

c) IAs appear to be more effective administratively than Public Irrigation Schemes. Farmers are personally interested in their own organisation and this sometimes permits them to do the same job cheaper and faster. Thus IAs reduce the need for a heavy public bureaucracy to run the irrigation schemes. The civil servants utilised in the development of an irrigation scheme can be moved to a new scheme when the first is completed, instead of remaining as permanent employees of the developed scheme.

d) The recovery of water fees is more effectively carried out through IAs than through Public Irrigation Schemes. Experience shows that payment of water fees is higher among farmers of IAs than in those run by government officials.

e) Rules and regulations are respected better since the punishment of faults in IAs is effective and quick. Most of the IAs have their own irrigation juries or regulations that are highly respected by members, so making any punishment effective.

f) The relations between those distributing the water (water masters and water guards) and the farmers receiving it are much more friendly and cooperative than when the same job is undertaken by personnel of the administration. In the latter case, the water masters and ditchriders are seen as representatives of the government and not as part of the farmers' group.

g) The IAs provide an excellent communication channel between the administration and the farmers. When they do not exist, communication tends to be one-sided from the top to the bottom.

h) The Board of Directors and often the appointed Manager himself are farmers of the irrigation scheme and continuation in their jobs will depend largely on their performance. This is indeed an important incentive to carry out their jobs in the best possible way.

In spite of the fact that IAs are a highly desirable way of managing an irrigation scheme, they also have some limitations:

1) The water distribution system utilised frequently leads to considerable operational water losses. Most of the IAs operate on the basis of a 'semi-demand' system, i.e. the farmer requests water and within a period of 3 to 7 days his request is granted. Although this system is perhaps the most desirable one from a social point of view, it is less efficient than the rotational system. This can be easily understood when one considers that to grant the request of a few farmers sometimes the whole canal irrigation network has to be kept in operation, with the corresponding water losses.
2) Sometimes old water rights are an obstacle to efficient water distribution. In the course of their history many old irrigation associations have developed particular water rights which are incompatible with an efficient operation of the network. Such systems usually grant a fixed amount of water to the landowner and such a right can be sold or bought from others. This leads to all kinds of transactions—many of which bring social unrest—and subsequently to the need for constructing small but long ditches to bring the water to the buyer's land. Operational losses are bound to be high in this kind of water distribution system and consequently infiltration losses will be considerable owing to the complexity of the distribution.
3) There is little capability for undertaking responsibilities outside the operation and maintenance field. Most of the people running an IA are from the

farmers' community and are unlikely to have technical and managerial skills, thus tending to limit their sphere of competence to problems that they themselves can easily handle. So the involvement of IAs in farmers' training activities or water research investigations, monitoring of water quality problems, marketing and agricultural processing activities, etc. becomes problematic. However, it is possible for the IA to become more involved in activities other than those purely related to operation and maintenance problems by having qualified personnel. The Irrigation and Water Districts of the USA prove that under proper technical guidance and with sufficient financial resources, IAs can be very dynamic organisations with a wide sphere of competence.

4) Long periods of considerable effort are required to get an IA established and working properly, even in countries like Spain where such institutions have centuries of tradition. The farmers"' irrigation experience, educational level, the tradition of such associations in the region, the absence or presence of farmers' leaders are all factors that may influence the time required for the proper functioning of such organisations. In favourable conditions the IA can be operational a few years (5 or less) after completion of the irrigation works, while in less favourable conditions long periods of 15-20 years may be required.

iv) *Size of the Association:* The question of the optimum size of the Association is important in determining its potential capability to undertake the related functions efficiently. Indeed, up to a certain size of irrigation scheme (about 5000 ha), the size of the Association is predetermined by the physical size of the scheme. In larger schemes, the possibility of subdividing them into irrigation sections, each having its own IA, or having a large association for the whole scheme, is frequently debated.

Large IAs have the advantages of greater economic potential and negotiating power. However, the larger the IA the more difficult becomes communication between the individual farmers and the executive body. In countries where the means of communication (mainly telephone) is effective and widespread in the rural areas, the formation of large IAs may be desirable (this is the case in the Irrigation and Water Districts of the USA), otherwise a federation of smaller associations may be preferable.

On the other hand, small IAs facilitate communications; but the administrative costs are greater and therefore place more of a burden on the farmers.

From these discussions it appears that the optimum size for an irrigation association is neither too large nor too small: a size corresponding to 10000 ha (2000 - 4000 farmers) is perhaps a desirable target. This should be matched with the criterion that each IA should operate within the command area of hydraulically independent canals, as far as possible.

Public Irrigation Schemes

Public Irrigation Schemes (PIS) are mainly controlled by government officials, in contrast with the Irrigation Associations which are predominantly controlled by the farmers. In some countries these organisations are of a permanent nature while in others they are established to prepare the way for transferring responsibilities to the farmers.

These institutions are, in general, closely connected with the organisation used during the construction stage of the irrigation project. Frequently, most of the executive and administrative personnel employed during the construction period remain to run the irrigation scheme after completion of the works. The risk involved is that too little attention may be paid to agricultural problems.

i) *The structure:* The main executing bodies of the Public Irrigation Schemes are:

a) *The Manager's Office:* The Manager is in charge of the implementation of the plan previously approved by the central office which may be the Ministry of Agriculture, the Ministry of Irrigation or the Ministry of Public Works. An able manager is crucial to securing successful scheme administration. Sometimes the manager is supervised by a regional committee comprising representatives of several ministries. This is the case when emphasis is placed on regional development. He is also sometimes assisted by a group of advisers.

b) *The Executive Bodies:* Are the operational, maintenance and administrative units. The organisation is characterised by a clear line of command that goes from the national level to the lowest units.

The case in which a PIS takes on further responsibilities will be discussed later under the so-called Multipurpose Public Irrigation Schemes. However, even in cases where a PIS does not take on direct responsibility for the provision of such additional services, it can play an important role in ensuring that such services are effectively provided for the scheme. For this purpose a supervisory board could be established where representatives of the institutions responsible for providing the services are members of the Board together with the farmers' representatives.

ii) *Main characteristics:* The main distinctive features of these organisations are:

a) They offer good possibilities to rationalise water distribution. They are not bound by obsolete water rights and can enforce the type of water distribution considered most efficient. However, it has to be recognised that such possibilities are frequently disregarded and water distribution is not much better than with IAs.

b) Due to the presence of technical personnel in these institutions, a considerable amount of technical data is kept which is extremely useful to plan the future O&M of the scheme properly.

c) When sophisticated techniques are used for the distribution of water, such as dynamic regulation of canals, distribution on demand, etc., the use of these institutions is almost inevitable since the management of such systems is beyond the capacity of farmers' organisations.

d) The complex operation and maintenance of very large canals, especially those having multipurpose use, may also be justified since this may be the only way to coordinate conflicting interests.

e) Little time is required to get the institution established.

f) Where farmers have little or no experience in irrigation, the PIS may be the only way of running the irrigation scheme during the initial stages, until the farmers learn enough about the scheme to handle it themselves. However, there is the risk that these institutions can tend to be self-perpetuating unless otherwise specified by law.

g) To function properly, a PIS needs strong institutional support at the national level. In other words, these organisations should be part of an autonomous or national institution dealing with the management of irrigation schemes.

h) The manager of the irrigation scheme and some of the professionals in charge of the different units are generally public officials, who often stay for a few years in the project and are then transferred to other projects. This lack of continuity of staff does not provide incentives to improve their performance in their jobs.

i) The project staff are not accountable to the farmers—who are the most interested in the efficient running of the system—but to the central administration. In very large bureaucracies efficient financial control is often not possible, especially where it is considered that part of the funds comes from two sources: the water fees paid by the farmers and the government subventions. Under these circumstances misuse of funds often occurs.

Specialised Water Management Organisations

The establishment of Public Irrigation Schemes (PIS) is relatively simple since it only requires some qualified personnel, equipment and financing, elements which are all generally present during the construction of the irrigation system, and therefore it is not normally difficult to guarantee continuity during the subsequent operational phase.

The question is much more complicated when the intention is to transfer the management of the scheme to the farmers, in particular when trying to establish IAs. However, the effort is worth it. Most researchers agree that the role of an IA can be instrumental in increasing irrigated agricultural production in many countries, and any expenditure made in this direction gains high returns.

The main problem in establishing IAs lies in the difficulty of explaining to the future members of the association what an association is, and then in convincing them of the convenience of its establishment. Furthermore, there is also the problem of training a small group (the executive board) in operation and maintenance matters in such a way that they will be able to undertake these functions properly.

There are two main approaches that can be followed by countries to overcome these problems, referred to as:

— the model approach

— the training approach.

i) *The model approach:* The idea behind this approach is that farmers learn mainly by seeing. In the initial year of operation of the irrigation scheme, a PIS is set up which then serves as a model to show how the system should be managed. During these early years, a proposal is made that an IA should be established. When the IA is prepared to take on the responsibilities, the corresponding functions are transferred. This is usually a long process and experience in Spain, where such associations have a long tradition, shows that an average of 10-15 years is needed before the IA is properly established.

This approach has the great advantage of permitting an almost immediate utilisation of the water distribution system since the establishment of a PIS is a relatively simple matter. It also has the advantage of allowing for a certain degree of "on-the-job" training. On the other hand, it has the great risk that once a PIS becomes settled, it may tend to perpetuate itself, thus delaying the transfer to the IA.

In spite of this latter risk, which can be avoided by appropriate laws, this approach has been found practical and is the one most commonly followed.

ii) The training approach: Where farmers already have some notion about irrigation practices and a good attitude towards cooperation exists, the training approach can be tried out. Under these conditions it is possible to give to farmers, and especially to their leaders, some short training that permits early establishment of the IA. The Philippines is adopting this approach through its Farm System Development Corporation (FSDC) and the programme is quite successful. However, this approach requires strong support from the government—the FSDC was established by Presidential Decree in 1975—and well-qualified and specially trained groups of people capable of undertaking the training of the farmers' leaders.

Irrigation Schemes with Mixed Control

This designation is applied to irrigation schemes where part of the physical system is handled by government officials and the rest by the farmers. This management system is becoming increasingly popular in many countries of the Far East and Asia where water distribution problems in the watercourses have always been acute. The distribution of water in the watercourses has often been a no-man's land due to the lack of definition of responsibilities for their operation. Furthermore, it is practically impossible for the few government officials to handle the distribution of water when large numbers of farmers are involved. This has led to some innovative actions where the government is promoting the formation of irrigation associations for each Watercourse or tertiary canal, while the rest of the irrigation network is still managed by government officials. This management alternative appears promising and its effect on some pilot projects has resulted in dramatic increases in crop production due to a more equal water distribution. Nevertheless, most of these experiences are not yet consolidated and their evolution should be monitored in order to learn more about their potential.

Regarding the main system managed by government officials, all that has been said for PIS applies. The crucial point remains the relationship between the associations and the officials managing the main system. It is most likely that most of these associations will evolve into a federation that will have a strong voice in all the management affairs of the scheme.

Multipurpose Water Management Organisations

A multipurpose water management organisation is characterised by having responsibilities directly related to water management and some others which may include irrigation extension, research, water quality management, marketing, etc. These responsibilities should be the result of a careful examination of what is needed in the project that cannot be effectively provided by existing institutions.

In many cases a fully integrated management organisation would be unnecessarily complex and would lead to duplication of existing institutions, while the use of specialised water management organisations would not provide sufficient services. In this case the multipurpose water management organisation represents a suitable solution since it uses the nucleus of a water management organisation and adds only those activities that are most needed and are not done by any other existing institution.

This kind of organisation is unusual among irrigation associations because there are several obstacles to establishing one. First of all, the water laws of many countries state categorically that the functions of any irrigation association must be limited to aspects related to the management of irrigation water, leaving no room for the addition of any other activity. Secondly, farmers' associations have to live from the contributions of their members, and it is unlikely that they will dedicate any effort to activities which do not appear financially attractive such as irrigation extension and/or applied research. Thirdly, most of the irrigation associations have traditions established long ago and are somewhat reluctant to undertake any additional responsibility.

Therefore, multipurpose organisations basically take the structure of a PIS where the relevant new activities are added. They have been designated here as Multi-purpose Public Irrigation Schemes (MPIS). Typical added functions are: irrigation assistance to farmers, applied irrigation research, irrigation improvement service, water quality management, etc. Sometimes a MPIS may also take responsibility for the establishment of some physical infrastructure aimed at improving marketing (stores, processing plants, markets), provision of inputs (fertiliser, seeds, etc.) or credit facilities, depending on the specific needs of the project area. While the establishment of the first group of functions (formative and training nature) is strongly favoured in many instances, the second group must be restricted to very specific situations, and only if they cannot be undertaken in the near future by any existing institution.

Of particular importance is the assistance to farmers in matters related to irrigation practices. In many circumstances, the establishment of an irrigation assistance service within the project organisational structure can be fully justified and effective. Another important responsibility which is often added in irrigation schemes established on newly developed land relates to applied research. In such circumstances, it is most convenient to gain experience rapidly in the management of soils and new crops. For this purpose, applied research farms are established by the MPIS, which can also be very useful for demonstrations of good agricultural practices to the farmers. If such farms are properly managed, the income from the agricultural produce often covers all the research expenses and therefore it is not a financial burden in the management of the scheme.

Most of the characteristics described for PIS also apply to this multipurpose type, although the latter has the great advantage of providing some supporting services (irrigation assistance, research, etc.), which are nearly indispensable in the early stages of development of an irrigation scheme. It is felt that the provision of these services—in one way or the other—is the key to achieve proper water management at the farm level.

References

Jones,W., "The World Bank and Irrigation", *A World Bank Operations Evaluation Study*, OED, World Bank, 1995.

Langford, K., Forster, C. & Malcom, D., "Toward a Financially Sustainable Irrigation System, Lessons from the State of Victoria, Australia, 1984-1994", *World Bank Technical Paper*, 1999.

International Commission on Irrigation and Drainage (ICID), "The Watsave Scenario", ICID, New Delhi, 1998.

______________, "ICID Strategy for Implementing the Sector Vision of Water for Food and Rural Development", ICID New Delhi, 2000.

15

Irrigation Challenges

According to the International Water Management Institute (IWMI) reliable and timely delivery is the exception, not the rule, in most developing country irrigation systems. Typically farmers in the more favoured parts of irrigation systems receive an adequate supply, while those at the tail end can face ruin. Poor management directly aggravates existing inequities within the farming community. Many farmers, frustrated by unreliable water deliveries, have opted to install tubewells. Consequently, groundwater use is exploding in many surface systems to compensate for the unreliable service from canal systems. Perhaps it will require a shock such as occurred in the 1960s and 1970s to awaken policy makers and the public at large to take the necessary actions to improve the management of these canal systems.

Lack of Irrigation Investments

Expenditures for agriculture over the last decade has fallen significantly due to a number of factors. Most frequently cited are:

— *The potential for expansion of irrigated area remains limited:* the water resources of many rivers have been fully exploited. The competition for water resources between agriculture and the demands of an increasing urban population and more stringent environmental rules and regulations reduce the potential for expansion of irrigation.

- World food prices are close to historical lows, reinforcing a perception of food sufficiency. The economics of irrigation development is less favourable than in the past two decades.
- The unit cost of development increased as the most suitable sites were exploited. Irrigation construction costs have risen to two to three times their previous level.
- Most investment is now for the rehabilitation of existing schemes. The cost of rehabilitation per hectare is thought to be considerably less than for a new project.

Less well known but equally important are adverse administrative and behavioural reasons:

- Environmental concerns have discouraged investments in irrigation and dam construction, particularly among the international development banks.
- Donors are responding to the new broad agendas of their organisations focusing on poverty alleviation, privatisation and environment. Private and public sector development, health and education now receive more aid than traditional infrastructure.
- Budget cuts for project identification, preparation and supervision contribute to the shift away from costly and time-consuming irrigation projects.

The perception of donors, supported by the international community of water resources and irrigation experts, is that the main cause of the poor performance of irrigated agriculture in many developing countries are deficiencies in management, institutions and policies, not the technology. This perception supports the view that fewer investments are needed for irrigation.

The consequence of declining investments for irrigated agriculture is a decline in rate of growth of irrigated areas and a backlog in rehabilitation and modernisation of existing systems. A notable exception to the overall decline in irrigation investment is Saudi Arabia. However, the excessive use of irrigation water has resulted in the rise of shallow groundwater in some areas with negative impacts on the environment, human health and engineering facilities.

Delayed Groundwater Development

Over time, the area irrigated by groundwater has increased in importance around the world. Groundwater development has been growing at an exceptional rate in recent decades. More reliable water delivery and declining extraction costs due to advances in technology and, in many instances, government subsidies for power and pump installation have encourages private investment in tubewells.

According to a report of the World Commission for Water, aquifers are being mined at an unprecedented rate. About 10 per cent of the world's agricultural food production depends on using mined groundwater. Water tables are dropping in fossil aquifers, including in the Western Unites States.

The recent development of groundwater has led to the over-exploitation of groundwater resources in some semi-arid regions where water tables have been falling at an alarming rate - often one to three metres a year. These regions include some of the world's major grain producing areas such as the Punjab and the North China plain. Groundwater development for irrigation is more recent in some countries. In Thailand, the explosive use of diesel pumps in the Chao Phraya and Mekong projects has responded to the increased demand for dry season cultivation of high-value crops and the unreliable supply from these large gravity irrigation systems employed.

The rapid development in groundwater, however, comes at a price. In many arid areas, more and more areas are in overdraft and the associated water-quality problems are emerging. In India, unreliable power supplies combined with weak management of groundwater resources greatly constrain the growth of irrigated agriculture. The focus of the world's attention on the environmental degradation and dislocation associated with dams has obscured another major environmental issue in the management of water resources.

The explosion of the use of pumps for irrigation, domestic and industrial use is degrading groundwater resources. The point has been reached in some areas that the over-exploitation is posing a major threat to the environment, health and food security.

The limit of groundwater development for irrigation in the world's major 'breadbasket' countries may be reached in a few years - before the end of the present decade in India. Over-exploitation of groundwater resources in arid and semi-arid countries has been documented in recent studies - Ogallala aquifer in the United States, Coastal aquifer and Souss aquifers in Morocco, Hermosillo in Mexico. It is expected that the development of groundwater for irrigated agriculture is going to slow down in the coming years.

Adoption of Water Conservation Application Techniques

A number of farm application irrigation technologies that could contribute to the conservation of water in both surface and piped systems have been developed. These techniques are generally adopted for the cultivation of high value crops. About 20 million hectares worldwide are now irrigated by sprinkler irrigation. Micro-irrigation as a technology has matured into reliable water and fertigation management system for crop production over the past decades and its usage continues to increase rapidly. A total of 3.2 million hectares are being irrigated by micro-irrigation techniques throughout the world, which represent one per cent of the total areas irrigated in the world. A growth of more than six times over the last 20 years has been achieved, which is quite a success. Micro-irrigation is an important tool in the drive towards food security. Micro-irrigation systems have great potential to create significant opportunities for smallholder agriculture.

Advanced surface irrigation techniques such as precise land levelling, gated pipes or cablegation have also been developed to improve the typical poorly distributed application of water under conventional methods by furrow, border or basin irrigation. Experience shows that these techniques are too sophisticated to be adopted by most smallholders in developing countries. Adaptive research in the Middle East and North Africa has shown the difficulties in introducing surface water conservation techniques. Many attempts to introduce gated pipes in Morocco for example have failed for a number of reasons, mainly the poor maintenance of the quaternary earth canals. Considerable efforts for training of farmers in water conservation in surface systems are needed.

Farmers have adopted pressurised systems much faster than surface application techniques. For example, about 65 per cent of the farmers in the Jordan valley have shifted from surface to drip irrigation over a 10-year period. Farmers built on-farm storage reservoirs to provide the flexibility required for drip irrigation.

Issue of Technology vis-à-vis Management

The opinions differ among experts on the causes of poor performance of canal systems. The following statement epitomises the thinking of many irrigation professionals:

> "Irrigation schemes in many parts of the world are known to be performing well below their full potential... There is now wide recognition that deficiencies in management and related institutional problems, rather than technology of irrigation, were the chief constraints of poor performance of irrigation systems".

There is no question that management of irrigation systems has been haunted by a multitude of problems. Admittedly there are some important management-related and institutional deficiencies, such as conflicts between farmers and irrigation agencies, farmer interference and lack of discipline, poor coordination between government agencies and poor recovery of investments and recovery costs and poor farmer incentives. Few irrigation experts have challenged the widespread wisdom that these are the main causes of poor performance of irrigation systems.

The following are the causes of deficiencies in designing irrigation systems

- Most civil engineers are well trained in structural engineering and construction techniques but not in the practical and theoretical aspects of unsteady flow hydraulics that are the norm in most irrigation systems. They are also unfamiliar with the constraints of the end user - i.e. on-farm irrigation management requirements. Appropriate irrigation design and management is much more complicated than most engineers, administrators and donors assume.
- Second, designers are rarely confronted with the consequences of how their designs function once they are installed.
- Third, many irrigation agencies cling to outdated design standards and often resist changes by external experts. Most consulting firms have no contractual motivation and no financial incentives to introduce new concepts.
- Operational failures of irrigation systems are not dramatic and are not widely publicised. Operation staff can operate canal for a while by infringing on freeboard and farmers are adjusting to the poor delivery by sinking wells or building reservoirs.

References

Plusquellec, H., "How design, Management and Policy Affect the Performance of Irrigation Systems: Emerging Modernisation Procedures and design Standards", *FAO*, Bangkok, Thailand, 2002.

Reinders, F., "Micro-irrigation: A World Overview", *Proc. Sixth International Micro-Irrigation Congress*, Capetown, South Africa, 22-27 Oct. 2000.

Shah, M. & Strong, M., "Food in the 21st Century: From Science to Sustainable Agriculture", CGIAR Secretariat, World Bank, 2000.

Bibliography

Abdi Qassim, *Sprinkler Irrigation, Situation Analyse,* IPTRID, 2003.

Addink, *et al., Design and operation of farm irrigation systems,* ASAE, 1989.

Bauder, J.W., *Simple Yield Models Based on Water,* Montana State University, Bozeman, Personal communication, 1998.

Collier, R.L. and Rochester, E.W., *Water Application Uniformity of Hose Towed Traveller Irrigators,* ASAE Transactions, 1980.

Dadiao, C. and W.W. Wallender, "Droplet size distribution and water application with low-pressure sprinklers", *Transactions of the ASAE,* 1985.

FAO, "Mechanized Sprinkler Irrigation", *FAO Irrigation and Drainage Paper,* 1982.

Finney C E., "The agricultural benefits of sub-surface drainage on irrigation schemes", *ICID Journal,* New Delhi, India, 1997.

Frost, K.R. and H.C. Schwalen, "Evapotranspiration during sprinkler irrigation", *Transactions of the ASAE,* 1960.

Hill, R.W. and R.G. Allen, "Simple Irrigation Scheduling Calendars", *Journal Irrigation and Drainage Division,* ASCE. 122:2, March/April, 1996.

Hill, R.W., "Consumptive Use of Irrigated Crops in Utah", *Utah Agr. Exp. Stn. Res. Report No. 145,* Utah State University, Logan, UT. Oct., 1994.

Hoffman, G.J., Howell, T.A. and Solomon, K.H. (eds.), "Management of Farm Irrigation Systems", *ASAE,* St. Joseph, 1990.

Isrealson and Hansen, *Irrigation Principles and Practices,* 1967.

Johnson S H., *Irrigation management transfer: decentralising public irrigation,* IIMI, Mexico, 1993.

Kay, M., *Sprinkler Irrigation: Systems and Practice,* Bedford, 1986.

Keller, J. and Bliesner, R.D., *Sprinkler and Trickle Irrigation,* Chapman and Hall, New York, 1990.

Kincaid, D.C., M. Nabil, and J.R. Busch, "Spray losses and uniformity with low pressure center pivots", *ASAE Paper,* 1986.

Kincaid, D.C., *Spray drop kinetic energy from irrigation sprinklers,* 1996.

Mao, Z., "Environmental impact of water-saving irrigation for rice", In: *Irrigation Scheduling: From Theory to Practice, Proceedings ICID/ FAO Workshop,* Rome, Sept. 1995.

Myers, J.M., C.D. Baird and R.E. Choate, *Evaporation losses in sprinkler irrigation,* Florida Water Resources Research Center Publication No. 12. University of Florida, Gainesville, 1970.

Norum, E.M., "Sprinkler System Returns Wastes to the Land", *Published in World Farming magazine,* June, 1973.

Pair, C.H., W.H. Hinz, K.R. Frost, R.E. Sneed, and T.J. Schiltz, *Irrigation,* The Irrigation Association, Silver Spring, MD, 1983.

Reinders, F., "Micro-irrigation: A World Overview", *Proc. Sixth International Micro-Irrigation Congress,* Capetown, South Africa, 22-27 Oct. 2000.

Seginer, I., "Water losses during sprinkling", *Transactions of the ASAE,* 1971.

Shah, M. & Strong, M., "Food in the 21st Century: From Science to Sustainable Agriculture", CGIAR Secretariat, World Bank, 2000.

Skutsch J C., "Maintaining the value of irrigation and drainage projects", *Report OD/TN/90,* HR Wallingford, Wallingford, UK, 1998.

Small L E & Carruthers I D., *Farmer-financed irrigation: the economics of reform,* Cambridge University Press, Cambridge, UK, 1991.

Thomas F. Scherer at all, *Sprinkler Irrigation Systems,* MWPS, 1999.

Vermillion D L., "Impacts of irrigation management transfer: a review of the experience", *Research Report No.11,* IIMI, Colombo, Sri Lanka, 1997.

Walker, W.R. and Skogerboe, G.V., *Surface Irrigation, Theory and Practice,* Prentice Hall, Englewood Cliffs, New Jersey, 1987.

Withers and Vipond, *Irrigation Designs and Practices,* 1974.